Nanostructuring of Metals by Severe Plastic Deformation

By
Scotty Bown

1 Introduction

The present study aims to explore the melting dynamics and heat transfer in latent heat energy storage (LHES) systems by investigating the effects of heating surface geometries, enclosure configurations, periodic boundary conditions, and enclosure orientations on PCM performance. The research explores melting behaviour over flat, wavy, and arc-corrugated surfaces heated horizontally from the bottom, as well as wavy surfaces with varying amplitude-to wavelength ratios under periodic boundary conditions and roughened surfaces at different inclination angles. The fundamental objective is to enhance heat transfer performance and contribute to advancements in LHES technologies. This chapter provides an overview of Rayleigh-Bénard convection in PCMs and its applications, emphasizes the importance of analyzing melting dynamics, and highlights the role of surface geometry and variation in surface heat flux in augmenting melting and heat transfer characteristics.

1.1 Overview of R-B convection in PCM and applications

Phase change materials (PCMs) are extensively utilized in latent heat energy storage (LHES) systems owing to their high energy storage density and nearly isothermal phase transition behaviour. The effectiveness of such systems is largely governed by the rate of heat transfer within the PCM during both melting and solidification. Since most PCMs possess inherently low thermal conductivity, the melting process cannot be attributed solely to conduction process. Buoyancy-driven natural convection process develops within the molten region subsequent to the onset of instability as the liquid fraction increases and this significantly enhances the heat transfer and melting process. R-B convection arises when buoyancy forces generated by temperature gradients surpass the opposing effects of viscous resistance and thermal diffusion. This mechanism dictates the fluid motion and thermal transport within the molten zone, exerting a strong influence on the overall heat transfer efficiency of the latent heat storage system. Therefore, a comprehensive understanding of R-B convection is vital for designing and optimizing PCM-based thermal storage devices, thermal management units, and energy-efficient enclosures. The onset of convection is governed by the Rayleigh number (Ra) a dimensionless parameter expressing the ratio of buoyancy to viscous and diffusive forces. Convection commences when the Rayleigh number exceeds a critical value, typically $Ra_{cr}\approx1708$ for an infinite horizontal layer with rigid boundaries [2,3]. In PCM melting systems, the Rayleigh number evolves dynamically as the melting front advances, since the liquid layer thickness increases with time, thereby altering the characteristic length and the effective temperature gradient. The continuous interplay between conduction and convection renders the melting process transient and nonlinear. The intensity of convection and the associated heat transfer augmentation depends on several parameters, including:

- The temperature difference between the heated and cooled surfaces,
- The geometry and orientation of the enclosure,
- The surface configuration or roughness of the heating wall,
- The thermophysical properties of the PCM, and
- The boundary conditions imposed (such as isothermal, constant heat flux, or periodic heating).

At higher Rayleigh numbers, the convective flow transitions from steady, well-organized cells to unsteady or oscillatory behavior and, in extreme cases, to turbulence. Such transitions lead to further enhancement of heat transfer. Nevertheless, fully developed turbulence is uncommon in PCM applications due to the relatively modest temperature differences and moderate system dimensions involved.

Rayleigh-Benard (R-B) convection in phase change materials (PCMs) can be analysed through both numerical simulations and experimental visualization techniques. Simulation results typically show well-defined convection cells forming near the heated lower surface during the intermediate stages of melting. As melting advances, the number of plumes gradually decreases, giving rise to large-scale circulation patterns that dominate the flow field. Temperature contours often display steep gradients near the solid liquid interface at the onset of melting, which progressively become smoother as convection enhances mixing in the molten region. These visualization methods are valuable for identifying zones of strong heat transfer, stagnant regions, and the development of the mushy zone the partially melted layer where solid and liquid phases coexist. Figure 1.1 illustrates the three common regimes encountered during the melting of a PCM: conduction, linear, and coarsening. Initially, melting is conduction dominated, with heat transfer occurring mainly by conduction from the heated surface and negligible fluid motion in the liquid phase. As melting progresses and the liquid layer grows, buoyancy effects become significant, leading to the linear (convection-dominated) regime, where organized convective cells develop and the melt fraction increases almost linearly with time. In the final coarsening regime, these convective structures merge into fewer and larger cells, plume interactions weaken, thermal stratification becomes more pronounced, and the melting rate gradually decreases despite the continued presence of convection.

1.2 Importance of Assessment of Melting Dynamics

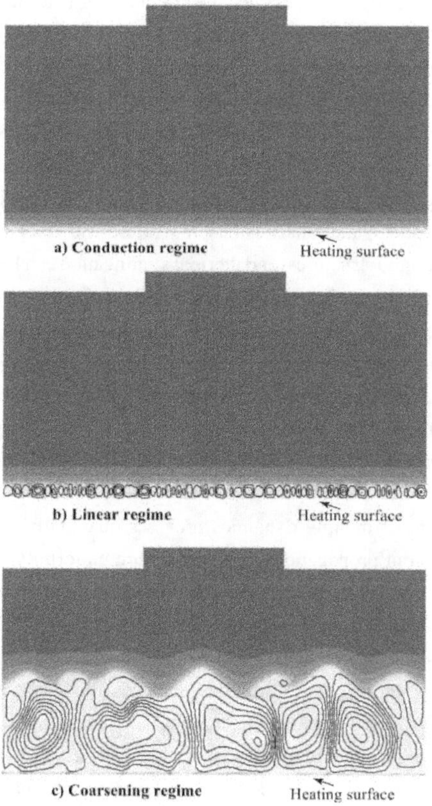

Figure 1.1: Regimes during melt progression

Phase change materials (PCMs) play a crucial role in thermal management, latent heat storage, and electronics cooling systems. Among the various phase change processes, melting the transition from solid to liquid is of particular interest in applications involving thermal energy storage [1] , heat exchangers [2] , solar thermal systems [3] , electronics cooling [4], and building energy systems [5]. The assessment of melting dynamics that comprising of the time-dependent evolution of temperature, phase interface movements, and flow characteristics within a PCM, is essential to understand and optimize the thermal performance of such systems. A comprehensive evaluation of melting dynamics enables better control of heat transfer mechanisms, improved design of enclosures, and enhanced efficiency of thermal systems.

The melting process is governed by a complex interplay of conduction, convection, and phase-change enthalpy. In the early stages of melting, heat conduction dominates near the heated surface, while as the melt fraction increases, natural convection within the liquid PCM accelerates heat transport. Assessing the melting dynamics allows researchers to capture these transitions between conduction-dominated and convection-dominated regimes. Through detailed temporal and spatial analysis of temperature and velocity fields, the development of convection cells, plume structures, and thermal stratification can be identified. This understanding helps in quantifying the Nusselt number variation with time and assessing the contribution of buoyancy-driven convection to overall heat transfer. Hence is vital to accurately evaluate how design modifications such as surface roughness, heating orientation, or boundary conditions-affect melting performance. Moreover, in systems with non-uniform or time-varying heating, such as oscillating thermal boundaries, the assessment of melting dynamics provides insight into unsteady thermal responses and the phase interface movement under periodic or transient heating.

The assessment of melting dynamics provides quantitative information on melt fraction evolution, liquid front propagation, and temperature uniformity within the PCM. A thorough understanding of melting dynamics helps identify thermal bottlenecks, such as stagnant zones or thermal stratification, which reduce effective energy utilization. For example, in large enclosures, the upper regions may remain unmelt due to weak convection currents, leading to incomplete energy storage. Assessing these patterns allows for better geometry optimization, such as adding fins, metal foams, or using multiple heat sources to improve uniform melting. Furthermore, accurate modelling and experimental assessment of melting dynamics enable cycle-to-cycle performance prediction. Over repeated thermal cycles, PCMs may undergo subcooling, phase segregation, or density driven stratification. Monitoring melting behaviour over time ensures the system maintains consistent thermal performance and helps in selecting PCMs with stable thermophysical properties for long-term use.

The configuration and nature of the heating surface strongly influence the melting rate and interface shape. For instance, heating from below promotes buoyancy induced convection and enhances melting, while side or top heating results in slower, conduction dominated melting. Similarly, the use of wavy, corrugated, or roughened surfaces can alter local heat transfer rates by promoting fluid mixing near the boundary.

Assessing melting dynamics enables detailed evaluation of how such geometrical or boundary modifications impact heat transfer and phase front evolution. Experimental and numerical studies often rely on time-resolved data of temperature contours, melt front shapes, and flow patterns to compare configurations and identify the most thermally efficient design.

In practical applications such as building-integrated PCM panels or heat exchangers this knowledge guides decisions on heater placement, inclination angle, and boundary condition selection (e.g., constant temperature vs. constant heat flux). Thus, the assessment of melting dynamics directly contributes to design optimization and performance improvement.

During melting, buoyancy forces induced by density variations generate complex natural convection currents within the liquid region. These currents give rise to plume formation, vortex motion, and circulation loops, which evolve dynamically as melting proceeds. The strength, size, and number of plumes significantly influence the overall heat transfer rate and interface morphology. By assessing melting dynamics through computational fluid dynamics (CFD) simulations or experimental visualization techniques (such as particle image velocimetry (PIV) and schlieren imaging), researchers can analyze how plumes originate, interact, and decay with time. Understanding these patterns allows for characterization of the flow instability mechanisms and transition from laminar to turbulent regimes within the melt region. Furthermore, quantifying plume behavior and thermal stratification provides valuable data for developing correlations between Rayleigh number (Ra) and Nusselt number (Nu), which are used in scaling and predictive models. Therefore, detailed assessment of melting dynamics bridges the gap between microscopic heat transfer mechanisms and macroscopic performance indicators. The solid-liquid interface is a critical region where heat transfer, phase change, and momentum interactions occur simultaneously. The shape and motion of this interface determine the effective thermal resistance and melt front stability. Assessing melting dynamics enables precise tracking of the interface evolution, allowing researchers to identify phenomena such as interface oscillations, mushy zone formation, and nonlinear melting fronts caused by surface curvature or nonuniform heating.

Comprehensive assessment of melting dynamics forms the foundation for developing validated mathematical and numerical models. Experimental measurements of melt fraction, temperature fields, and heat transfer coefficients are essential for verifying computational models such as the enthalpy porosity method used in CFD simulations. Accurate modeling requires understanding how thermophysical properties such as viscosity, density, specific heat, and latent heat vary with temperature and phase fraction. By comparing predicted and observed melting behaviors, researchers can refine model assumptions, enhance numerical stability, and ensure predictive accuracy.

This is particularly important for systems with complex geometries, multiphase flow, or unsteady heat sources, where analytical solutions are intractable. The combination of experimental assessment and numerical modeling leads to design guidelines and performance maps that are vital for scaling PCM systems from laboratory to industrial applications.

From an energy sustainability viewpoint, assessing melting dynamics is vital for optimizing energy storage efficiency, reducing thermal losses, and enhancing renewable energy utilization. Efficient melting ensures maximum latent heat absorption in minimum time, which directly improves the charge-discharge efficiency of storage units used in solar thermal power plants, waste heat recovery systems, and building envelope integration.

Understanding melting behavior also aids in material selection. By correlating melting rate and heat transfer characteristics with PCM properties, suitable materials can be chosen for specific thermal applications. For example, materials with high thermal conductivity and moderate melting temperatures are ideal for rapid response systems, while those with high latent heat and stability are better suited for long-duration storage. Thus, the assessment of melting dynamics supports global efforts toward sustainable energy systems by enabling the design of compact, efficient, and durable PCM based technologies.

Furthermore, the analysis of melting dynamics bridges experimental observations and theoretical models, leading to accurate predictive tools essential for engineering applications. Whether for energy storage, electronics cooling, or thermal regulation, understanding melting dynamics ensures efficient utilization of latent heat and supports the development of sustainable, high-performance thermal systems. The importance of assessing melting dynamics thus extends beyond academic curiosity it is central to advancing energy efficient technologies that meet the demands of a thermally conscious world.

1.3 Applications of evaluation of melting dynamics

Understanding R-B convection is crucial due to its wide-ranging applications in systems involving latent heat storage and thermal regulation. The major areas of application include:

Thermal Energy Storage Systems

In latent heat energy storage (LHES) systems, PCMs absorb and release thermal energy during phase transitions. The effectiveness of these systems relies on rapid melting and solidification. R-B convection enhances internal heat transport, thereby reducing charging time and improving heat release efficiency. Insight into convection behavior assists in designing optimized heating surfaces-such as wavy, corrugated, or finned geometries-that promote stronger buoyancy-driven flow and improve system performance [1].

Solar Thermal and Waste Heat Recovery

In solar thermal and waste heat recovery systems, PCMs serve as intermediate heat reservoirs to capture and store surplus thermal energy. The inherently nonuniform heating in such setups naturally triggers R-B convection within the PCM. Enhancing this convective flow through optimized surface geometry or enclosure configuration can significantly improve heat storage rates and operational stability, making the system more efficient and reliable [3].

Thermal Management of Electronics

PCMs are increasingly applied in the passive thermal management of electronic devices, batteries, and other high-power components. R-B convection within the PCM helps distribute heat more evenly, mitigating local hot spots and reducing temperature gradients. Controlling this convective process ensures consistent temperature regulation, extends component lifespan, and enhances the reliability of electronic systems under variable thermal loads [4].

Building Envelope Systems

In building energy applications, PCMs are integrated into walls, ceilings, and roofs to regulate indoor temperatures. They absorb excess solar heat during the day and release it during cooler periods, maintaining thermal comfort and improving energy efficiency. The R-B convection occurring within the PCM layer aids in uniform heat distribution, thereby enhancing the effectiveness of energy storage. Accurate modeling of this convective behavior is essential for predicting and optimizing the thermal performance of PCM-integrated building materials under varying environmental conditions. Despite its advantages, RB convection in PCMs introduces significant challenges in modeling and analysis. The continuously moving solid liquid interface, the time-dependent variation in liquid layer thickness, and the nonlinear coupling of flow and temperature fields make analytical solutions complex. To overcome these difficulties, researchers increasingly employ computational fluid dynamics (CFD) simulations to study melting and solidification under different boundary conditions, enclosure geometries, and heating modes. These numerical approaches provide detailed insights into transient convection patterns and have become indispensable for improving the design and performance of PCM-based thermal energy systems [5]

1.4 Importance of Surface Geometry in Melting and Heat Transfer Enhancement

Surface geometry plays a pivotal role in determining the rate and pattern of melting and the overall efficiency of heat transfer in systems involving phase change materials (PCMs). During melting, heat transfer occurs primarily through conduction at the solid–liquid interface and natural convection within the liquid phase. The configuration of the heating surface strongly influences these mechanisms by altering the temperature distribution, flow structure, and interface morphology. Therefore, understanding and optimizing surface geometry is essential for enhancing thermal performance and achieving uniform, accelerated melting in energy storage and heat exchange applications.

Melting initiates uniformly near the boundary in systems with flat heating surfaces, and heat transfer is dominated by conduction during the early stages. However, as the liquid layer thickens, buoyancy driven convection begins to develop. Although flat surfaces provide a simple and well-defined heat transfer path, they often lead to the formation of thermal stratification and stagnant zones, resulting in uneven melting and reduced energy efficiency. Consequently, surface modification techniques such as introducing roughness, waviness, or extended features are used to overcome these limitations. Wavy and corrugated surfaces enhance melting by increasing the effective heat transfer area and by disturbing the thermal boundary layer. The peaks and troughs on the surface act as local heat concentration points, which generate temperature gradients and promote stronger convective circulation within the molten region. This intensified mixing leads to a faster movement of the melting front and a more uniform temperature distribution.

Studies have shown that higher amplitude-to-wavelength ratios (AWR) in wavy surfaces can significantly accelerate melting at early stages of melting, although excessively large amplitudes may cause localized flow recirculation that slightly reduces overall efficiency. Similarly, rough or patterned surfaces such as those with square, triangular, or hemispherical protrusions can enhance convective activity by creating small vortices near the heated boundary. These micro-scale disturbances improve local heat transfer coefficients and delay the formation of stratified layers.

The shape and arrangement of roughness elements determine the strength and interaction of these flow structures. For example, square patterned tends to generate strong, well defined circulation zones, while triangular patterned promotes smoother flow transitions with moderate enhancement. As shown in Fig.1.2, on the sinusoidal wavy surface, the local surface inclination changes rapidly along the bottom. These frequent slope variations cause the buoyant rise of the melted fluid to originate from many slightly different directions. As a result, the upward motion organizes into several small, closely spaced buoyant upward plumes.Each plume generates a circulation cell, producing many small vortices.

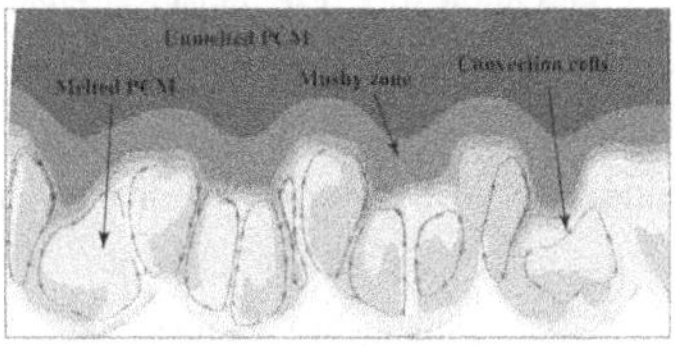

(**a**) Corrugated surface physics

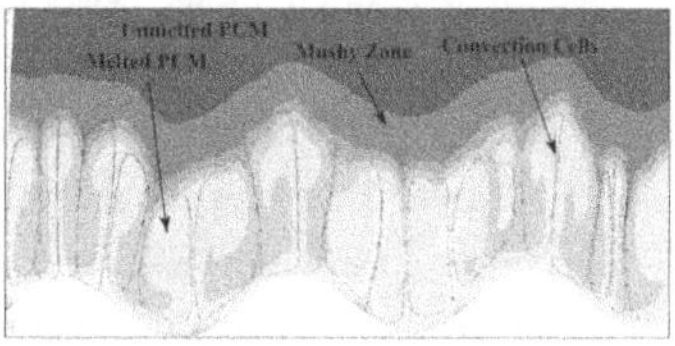

(**b**) Wavy surface physics

Comparison of physics involved in corrugated and wavy surface

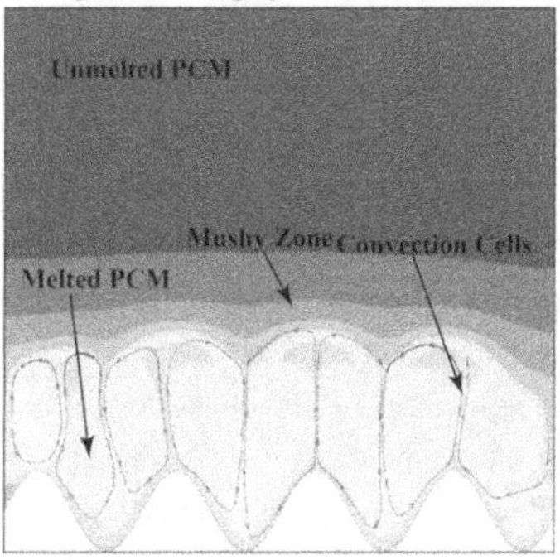

(**c**) Wavy surface A:λ = 2

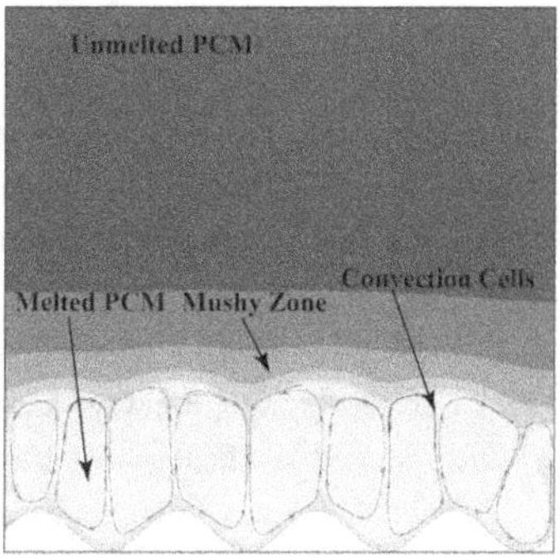

(**d**) Wavy surface A:λ = 1

Comparison of physics involved in wavy surfaces with different A:λ ratios

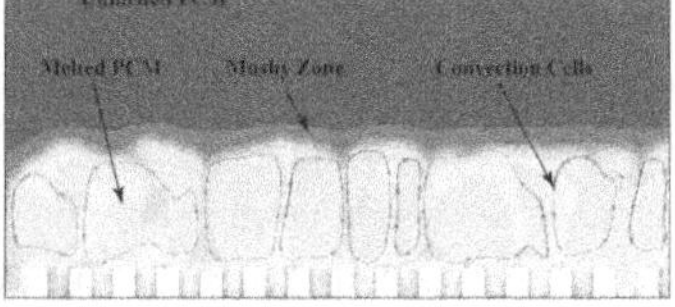

(**e**) Surface with Square patterned elements

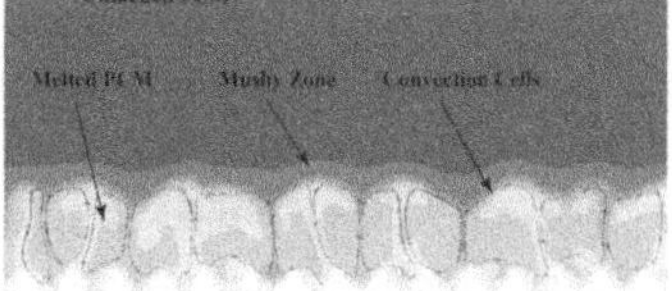

(**f**) Surface with triangular patterned elements

Comparison of physics involved in surface with square and triangular patterned elements

Figure 1.2: Importance of Surface Geometry in Melting and Heat Transfer Enhancement

1.5 Organisation of the thesis

The present work deals with exploration of the melting dynamics and heat transfer in latent heat energy storage systems. A detailed analysis of the effects of heating surface geometries, enclosure configurations, periodic boundary conditions, and enclosure orientations are carried out. Chapter 1 presents an overview of the study, outlining the background, applications of PCM, and the key characteristics of the wavy and patterned heating surfaces.

Chapter 2 reviews the recent experimental and numerical investigations on phase change materials. It discusses various experimental visualization methods and temperature measurement techniques, along with numerical approaches such as the Navier-Stokes formulation, the lattice Boltzmann method, and the enthalpy-porosity technique. The key findings from the literature, identification of the research gaps, and the objectives for the present study are also presented.

Chapter 3 describes the experimental methodology. It gives a detailed account of setup used to measure the melt fraction, temperature distribution, and temperature gradient visualization. It also details the numerical techniques employed to investigate melting dynamics and heat transfer in phase change materials, including the Navier-Stokes formulation with the enthalpy-porosity method, incorporation of temperature-dependent PCM properties, and validation of numerical results against experimental temperature and melt fraction data.

The analyses of flat, wavy and arc-corrugated heating surfaces in latent heat storage systems are explored in Chapter 4. Various surface geometries are analysed for their impact on melt fraction and time-varying temperature profiles, with flow patterns visualized via experimental and computational approaches. The study also lead to the development of the Nusselt-Rayleigh (Nu-Ra) correlations and non-dimensionalised thermal boundary layer thickness-Rayleigh (δ_T - Ra) correlations. Variation in boundary heat flux is key in real-life thermal equipment applications.

Chapter 5 examines the melting dynamics and heat transfer characteristics by modulating the amplitude-to-wavelength ratio and employing cyclic temperature heating. The influences on Nusselt-Rayleigh (Nu-Ra) correlations, liquid fraction, and melt height progression over non-dimensional time are analysed. These outcomes (Nu-Ra) are benchmarked against classical laminar scaling law [6] featuring the typical exponent 2/7 for Ra.

Chapter 6 investigates the impacts of patterned heating surfaces in conjunction with enclosure orientation on phase-change processes. A detailed exploration of the physics associated with melting dynamics of inclined PCM confinements is carried out. It analyses variations in melt fraction, temperature profiles, and thermal gradient distributions across diverse surface configurations. Additionally, the chapter presents comparisons of Nusselt-Rayleigh (Nu-Ra) correlations for various heating surfaces under different orientations.

Chapter 7 presents the salient conclusions derived from the thesis investigations and delineates avenues for future research endeavours.

2 Literature Review

This chapter presents a comprehensive review of research on phase change materials (PCMs) used in latent heat energy storage systems, solar thermal and waste heat recovery applications, and thermal management of electronic devices. It summarises the key experimental techniques adopted in previous studies and highlights major developments in numerical modelling approaches. The chapter also discusses the effects of various influencing parameters including enclosure size and shape, aspect ratio, surface roughness, cyclic (periodic) boundary conditions, and other design factors on PCM performance. Finally, the key insights drawn from the literature are outlined, followed by the specific objectives of the present study.

Introduction

Over the past several decades, research on phase change materials has largely centered on understanding flow behavior and enhancing thermal performance. Owing to the high costs associated with extensive experimental investigations, much of the work in this area has been carried out through numerical simulations. The existing studies can be broadly classified into two categories: experimental techniques and numerical modeling approaches. Among the numerical methods, the enthalpy-porosity technique [7] proposed by Voller and Prakash is one of the most widely adopted for simulating PCM melting. Experimental research, on the other hand, primarily focuses on measuring temperature distributions, melt fraction evolution, and visualizing flow patterns within the molten region.

With the continuous rise in global energy consumption, the need for effective thermal management solutions has become increasingly critical. Latent heat thermal energy storage (LHTES) systems incorporating phase change materials (PCMs) have gained significant attention due to their versatility and efficiency in a wide range of applications, including building temperature regulation [8], photovoltaic systems [9], cold storage, industrial waste heat recovery [10], aerospace, automotive systems, textiles [11], and cooling of electronic devices [12, 13]. Because these applications are highly sensitive to temperature variations, research on PCM-based thermal management has expanded considerably in recent years.

PCMs can undergo numerous reversible melting–solidification cycles, which improves sys-

---

A detailed presentation of the review has been published as
Sarath, K.P., Osman, M.F., Mukhesh, R., Manu, K.V. and Deepu, M., 2023. A review of the recent advances in the heat transfer physics in latent heat storage systems. Thermal Science and Engineering Progress, 42, p.101886. https://doi.org/10.1016/j.tsep.2023.101886

tem efficiency, enhances thermal stability, and reduces overall energy consumption ; therefore, they are widely used in thermal energy storage (TES) systems due to their high latent heat storage capacity.

2.1 Passive heat transfer enhancement in latent heat storages

Passive heat transfer enhancement in latent heat thermal energy storage (LHTES) systems involves improving heat transfer without external power input or mechanical devices. It relies on geometric modifications, material additives, or natural convection. Techniques include using fins, extended surfaces, metal foams, nanoparticles, and encapsulation structures.

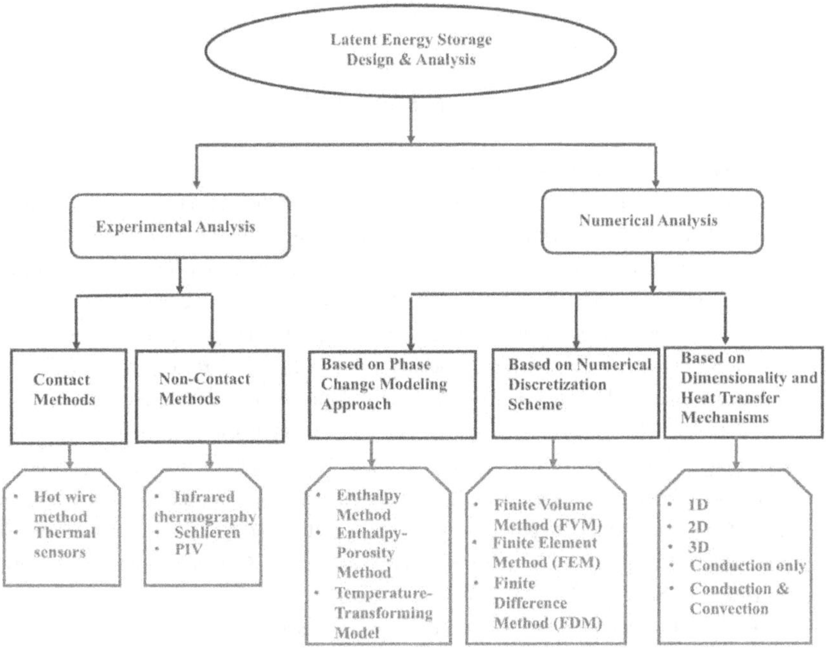

Figure 2.1: Methodologies for LES analysis

These methods increase thermal conductivity, enhance melting/solidification rates, and improve overall energy charging and discharging. Passive approaches are cost-effective, reliable,

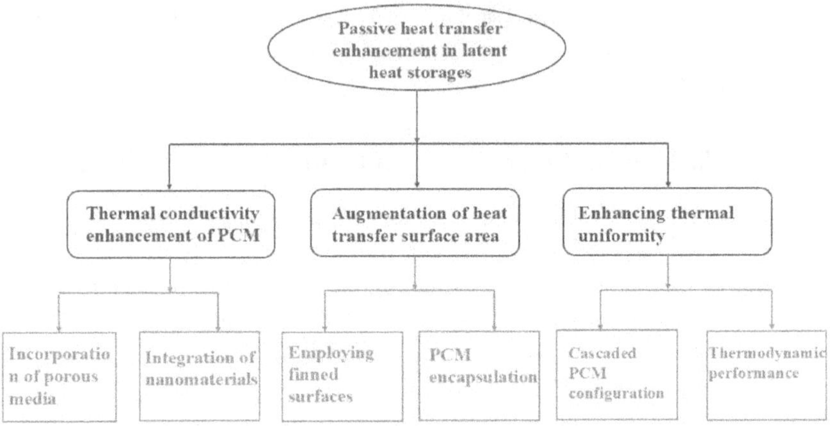

Figure 2.2: Passive heat enhancement methods in LES

and suitable for long-term thermal storage applications.

Despite their desirable ability to store large amounts of thermal energy, phase change materials (PCMs) suffer from a major limitation—their inherently low thermal conductivity—which restricts heat transfer during melting and solidification and ultimately reduces the efficiency of thermal storage systems. To overcome this drawback and improve heat transfer performance, two primary enhancement strategies are generally adopted. The first is a geometric modification approach, where the storage system is designed with enhanced surface area or integrated with conductive inserts such as fins, metal foams, honeycombs, or brushes to promote heat conduction and exchange. The second approach focuses on improving thermal conductivity by dispersing highly conductive nanostructures within the PCM. These include carbon-based nanomaterials (nanofibers, nanoplatelets, graphene flakes), carbon nanotubes, metallic nanoparticles (Ag, Al, Cu, C/Cu), metal oxide nanoparticles (Al_2O_3,, CuO, MgO, TiO_2), and silver nanowires. Incorporating these additives results in nano-enhanced PCMs (NePCMs) with significantly improved heat transfer characteristics.

Ho et al. [14] conducted melting experiments in a vertical square cavity using a solid–liquid phase-change material (n-octadecane) infused with Al_2O_3 nanoparticles, forming a nano-phase-change material (nano-PCM). Dhaidan et al. [15] experimentally and numerically investigated the melting of n-octadecane containing CuO nanoparticles within a square enclosure. Dhaidan et al. [16] investigate experimentally and numerically the melting behavior of nano-enhanced phase change materials (NePCM) in an annular cavity formed by two concentric cylinders. A numerical investigation of convection melting of water (ice)–copper nanoparticles as a nanoparticle-enhanced phase change material (NEPCM) in a bottom-heated rectangular cavity is conducted by Feng et al. [17]. Arasu et al. [18] conduct a numerical study to analyze the melting behavior of paraffin wax uniformly dispersed with Al_2O_3 nanoparticles inside a 25 mm × 25 mm square enclosure heated from one side. Ebrahimi et al. [19] perform a numerical analysis of the melting behavior of a NePCM within a square cavity configured with various arrangements of two pairs of embedded heat source–sink systems on its vertical sidewalls. A detailed numerical investigation is carried out to analyze heat transfer enhancement during the melting process in a two-dimensional square cavity through the dispersion of nanoparticles [20]. Wu, S. et al.[21] numerically investigates the melting behavior of Cu/paraffin nanofluid-based PCMs; for paraffin enhanced with 1 wt% Cu, the melting time is reduced by 13.1%. The numerical results show good agreement with experimental data in accurately describing the melting process. Jourabian et al. [22] found that nanoparticle-enhanced phase change material (NEPCM) improves thermal conductivity and reduces melting time when compared to standard PCM. Hosseinizadeh et al. investigate the melting of an RT27 in the presence of copper particles [23]. Simulation studies indicate that nanoparticles boost the thermal conductivity of NEPCM compared to standard PCM. Yanbin Cui et al. [24] experimentally study the melting of paraffin wax using carbon nanotubes (CNT) and fiber (CNF). Experiments demonstrate that adding CNF or CNT enhances the thermal conductivity of composite PCMs. Chibani et al.[25] investigated the melting process of phase change material inside a rectangular latent thermal energy storage unit using four nano-metal oxides (Al_2O_3, MgO, SiO_2, and SnO_2) with good thermal conductivity. The study found that nanomaterials greatly improve heat transport and solid melting rates. The type of particles did not significantly impact these outcomes.2.1

Incorporating high-conductivity fins or extended surfaces into latent heat thermal storage systems is a simple, reliable, and effective method for improving the melting rate. Gharebaghi and Sezai [26] examined the improvement in energy storage rate in a rectangular enclosure filled with paraffin (RT27) by incorporating partitioning fins. Huang et al.[27] conducted an analysis on the use of phase change material (PCM) as a method to limit the temperature rise within a silicon photovoltaic device, thus helping to maintain the photovoltaic module's electrical conversion efficiency. Shatikian et al. [28] numerically modeled the melting of paraffin wax PCM between vertical partitioning fins in a horizontal heat sink maintained at a constant base temperature. Akhilesh et al. [29] conducted a numerical investigation to determine the optimal size of a composite heat sink consisting of a vertical array of partitioning fins extending downward from a horizontal base plate. Levin et al. [30] proposed an optimization approach

Table 2.1: A summary of investigations on the melting behavior of NePCM within enclosed geometries

Author[ref]	Type of Analysis	Key findings
Ho et al. [14]	Experimental study – 25 mm by 25 mm square enclosure with Al_2O_3 nanoparticles.	The surface-averaged heat transfer results show that natural convection within the molten region decreases significantly as the nanoparticle mass fraction increases.
Dhaidan et al. [15]	Experimental and Numerical (FEM) study, square enclosure with CuO nanoparticles.	Adding nanoparticles enhances thermal conductivity, elevates temperature, and improves overall heat transfer, reducing charging time.
Dhaidan et al. [16]	Experimental and numerical study, annular cavity, CuO nanoparticles.	Adding more nanoparticles enhances melting performance by increasing effective thermal conductivity and strengthening natural convection.
Feng et al. [17]	Numerical (LBM) – Rectangular enclosure, Cu nanoparticle.	As the nanoparticle volume fraction increases, the temperature field and melting front develop more rapidly.
Arasu et al. [18]	Numerical, Al_2O_3 nanoparticles, square enclosure.	Examined the influence of heating surface orientation and volumetric concentration on melting performance.
Ebrahimi et al. [19]	Numerical, Al_2O_3 NePCM in a square cavity.	Observed that Case II yields the highest liquid fraction, while Case IV shows the lowest in final stages.
Sebti et al. [20]	Numerical, square enclosure, Cu nanoparticles.	Addition of nanoparticles improved heat transfer and led to a faster melting rate compared to the base PCM.
Wu et al.[21]	Numerical, Rectangular enclosure of 6 cm by 3.5 cm, Cu nanoparticles	The findings clearly indicate that adding nanoparticles significantly improves the phase change heat transfer of paraffin.

for designing a PCM-based heat sink intended for the transient cooling of electronic devices. Kamkari et al. [31] conducted an experimental study on the melting of phase change material (PCM) in a transparent rectangular enclosure, both with and without horizontal partial fins.

It is noteworthy that fluid flow and heat transfer over non-uniform surfaces frequently arise in various engineering contexts aimed at improving thermal performance, including microelectronics, solar flat-plate collectors, and condenser plates in refrigeration units. Undulating or textured surfaces may prove useful in phase-change material storage systems under conditions of prescribed wall heat flux.

A key advantage of such undulating or textured configurations lies in their ability to induce fluid agitation close to the surface, thereby yielding a higher overall heat transfer rate compared to a smooth flat surface. S. Khanna et al. [32] evaluated the optimal depth of a PCM container for different levels of solar irradiance and further investigated the influence of fin spacing, fin length, and fin thickness on this optimal depth. S. Liu et al. [33] examined the effect of various rectangular fin configurations on the melting performance of a PCM-filled vertical enclosure. Lu et al.[34] investigated the melting process of paraffin wax within a spiral finned tube. Experimental results show that spontaneous convection of melted paraffin has a significant impact on the paraffin's phase transition rate. Reddy [35] conducted a thermal modeling and study of a solar integrated collector storage (ICS) water heating system using transparent insulating materials (TIM) and phase-change material (PCM). The above literature work done with extented surfaces is summarised in Table 2.2.

The intricate shape of these surfaces generates correspondingly turbulent flow patterns in the adjacent fluid, which are governed by the nonlinear equations of the boundary layer. Research on the melting behavior of phase-change materials over sinusoidal undulating or textured walls remains sparse, covering only select heating scenarios and fluid types. The prevailing finding is that the aggregate heat transfer from any undulating or textured surface typically surpasses that from an equivalent flat surface, potentially varying with the amplitude-to-wavelength ratios of the profile. Conversely, the mean heat transfer coefficient per unit wetted area for such surfaces tends to be lower than for a flat plate. In heat exchanger design, however, the overall heat transfer rate holds greater practical significance. Kousksou et al. [36] performed a numerical investigation of melting dynamics in a rectangular sealed enclosure, with the lower undulating surface held at a constant temperature. In [37], researchers assessed the influence of wavy tubes on performance in a latent heat thermal energy storage unit featuring composite phase change material and metal foam, utilizing an enthalpy-porosity formulation for phase change heat transfer and solving it via the finite element method. Shahsavar et al. [38] investigated a novel triple-tube latent heat storage (LHS) system incorporating wavy channels

Table 2.2: A summary of investigations on melting behavior in enclosures with extended surfaces and fins

Author[ref]	Type of Analysis	Key findings
Gharebaghi et al.[26]	Numerical study, fin thickness (0.3 mm to 3 mm), height (5 mm and 10 mm).	Heat transfer improved as the fin spacing decreased in both the vertical and horizontal configurations.
Huang et al.[27]	Experimental, maximum 32 fins 4 mm spacing and thickness 0.5 mm.	The temperature rise on the front surface decreased, and thermal stratification within the PCM diminished as the number of fins increased.
Shatikian et al.[28]	Numerical 3D & 2D study, fin thickness (0.15 mm to 1.2 mm).	The melting rate increased as the fin spacing is reduced.
Akhilesh et al.[29]	Numerical 2D (FEM), varying height (0.03 m, 0.05 m & 0.07 m).	Neglected natural convection; concluded that increasing fins beyond a threshold does not yield significant improvement in thermal performance.
Levin et al. [30]	Numerical, minimum thickness of fins and gaps is 0.5 mm.	As fin length and interface heat flux decrease, temperature variation reduces, leading to an increase in the optimal PCM proportion.
Kamkari et al.[31]	Experimental, fins length 25 mm and thickness 4 mm.	Increasing the number of fins shortened melting time and enhanced total heat transfer, though the surface-averaged Nusselt number decreased.
S. Khanna et al.[32]	Numerical, fins thickness varied from 0 to 4 mm.	The results indicate that, for the selected parameters, the optimal PCM container depth is 2.8 cm at a solar irradiance of 3 $kWh/m^2/day$, while a depth of 4.6 cm is most suitable for a higher solar irradiance of 5 $kWh/m^2/day$.
S. Liu et al.[33]	Numerical, the fin had a thickness of 2 mm in the z-direction.	The results indicate that the optimal melting efficiency is achieved when two of the three fins are positioned in the lower half of the system.

Table 2.3: A summary of investigations on melting behavior in enclosures with wavy surfaces

Author[ref]	Type of Analysis	Key findings
Kousksou et al.[36]	Numerical study (FVM, 2D), Square enclosure (6 cm by 6 cm).	Results indicate that the melting rate rises with increasing amplitude of the wavy surface.
Ghalambaz et al.[37]	Numerical 3D study, wavy tube equation $r(z) = R_i - A(1 - \cos(2N\pi z))$.	The flat-surfaced tube delivers markedly superior performance relative to the wavy tube.
Shahsavar et al.[38]	Numerical 3D (FVM), wavy tube equation $s(y) = a_w \sin(2\pi y/L_w)$.	Wavy channels boost heat storage and recovery rates compared to straight walls.
Nematpour-Keshteli et al.[39]	Numerical 3D (FVM), wavy surface equation $y(x) = a\cos(2\pi x/L_w)$.	The wavy wall configuration and Y-shaped fins yield shorter melting times than straight walls.

with variable wavelengths during charging and discharging processes. Nematpour Keshtel et al. [39] proposed a strategy to enhance the thermal performance of paraffin as a phase change material (PCM) in solar flat-plate collector systems for residential and industrial applications, utilizing wavy surfaces integrated with Y-shaped fins and nanoparticles. Table 2.3 summarizes previous work on wavy surfaces. Experimental investigation of the melting behavior of phase change materials is essential for gaining insight into the flow and thermal dynamics within the liquid region. With recent developments in non-contact flow visualization techniques, it is now possible to monitor the progression of the melt front and capture the evolving velocity field more accurately. Moreover, precise measurements are crucial for validating numerical models and assessing the reliability of their predictions. Therefore, experimental examination of the phase-change process remains indispensable and warrants detailed discussion.

2.2 Advances in experimental analysis of melting of PCM

Table 2.4: A summary of recent experimental methodologies employed to study PCM melting

Author[ref]	Technique used	Key findings
Mukhesh et al. [40]	12 K-type thermocouples above the HTF passage, 6 beneath it.	Limited plume development above the wavy HTF passage improves melting performance and increases average melt height.
Dhaidan et al. [41]	5–6 thermocouples for each configuration (CF, CPF, and HE).	Heat exchangers with perforated fins exhibited enhanced thermal characteristics relative to those without perforations.
Lu et al. [34]	18 thermoresistors installed concentrically around the HTF and at the inlet/outlet.	Increasing inlet HTF temperature and flow rate significantly enhances heat transfer between HTF and PCM.
Yang et al. [42]	Differential Scanning Calorimetry (DSC) for thermal properties.	HTF temperature has a greater impact on the charging process compared to the HTF flow rate.
Lamberg et al. [43]	DSC testing with Mettler TA4000 and liquid nitrogen cooling.	Temperature distributions during melting and freezing are well estimated by both numerical methods.
Amirhossein et al. [44]	DSC on a Mettler Toledo Star for Oleic acid thermal properties.	OA placed above ice with a 50% volumetric ratio melts faster than all other binary PCM combinations.
Jevnikar et al. [45]	2D velocity fields measured using Particle Image Velocimetry (PIV).	Fluid velocity significantly affects heat transfer in the liquid domain, governing the melting pattern.

Experimental investigations are broadly categorised into two types: contact-based and non-contact methods. Earlier studies primarily relied on contact-based measurement techniques, which, despite accurately detecting the required variables, often interfered with and altered the natural flow characteristics. In later years, research shifted toward non-contact approaches, allowing the flow field to be observed without disturbing its natural evolution. However, the technical complexity and limited accessibility of these advanced methods still make them less commonly used for studying highly complex thermal and fluid behaviours. Further classification is based on the time-dependent behaviour of the measured parameters, dividing studies into steady and transient analyses. In steady-state conditions, convection currents stabilise and form a consistent temperature distribution across the enclosure. Conversely, transient studies capture the dynamic evolution of melting or solidification, where flow structures and thermal gradients continually change. Such transient evaluations are particularly important, as they provide deeper insight into the development of convection patterns and phase-change progression over time. This section provides a comprehensive review of contemporary studies utilising non-contact, time-resolved measurement approaches, which offer enhanced understanding of phase-transition phenomena. Progress in these techniques has enabled the acquisition of flow parameters with heightened accuracy, thereby enhancing the fidelity of computational models and simulations. Differential Scanning Calorimetry (DSC) stands as a prevalent method for probing phase-transition properties, especially in nano-augmented phase change materials (PCMs) [46]. While DSC adeptly elucidates the thermophysical responses of materials, it falls short in elucidating convective patterns or spatiotemporal temperature distributions. In contrast, Particle Image Velocimetry (PIV) has gained prominence as a robust methodology for mapping and measuring velocity distributions in fluidic contexts. For example, Liu [47] employed PIV to scrutinize convective dynamics and evaluate the performance of PCM enclosures partially infused with metallic foams. A curated compilation of pivotal recent experimental investigations on phase transitions is outlined in Table2.4.

2.3 Theoretical and Numerical Models for Heat Transfer in Phase-Changing Materials

A consolidated review by Hu and Argyropoulos [56] provides a comprehensive overview of the various modelling approaches used to address phase-change problems in PCMs. Building on this foundation, Heim [57] evaluated different formulations for representing sensible and latent heat transfer during phase transitions in building components. His results indicate that the Effective Heat Capacity Method and the Additional Heat Source Method produce only minor differences in predicted thermal behaviour, demonstrating that both approaches are adequate for most practical applications. Further advancement of theoretical modelling is reflected in the work of Javierre [58], who analysed two classical Stefan problems—melting and solid-state phase transitions—in binary metallic alloys. Their findings emphasize how interface evolution is strongly influenced by material composition and thermophysical properties. Additionally, Vogel and Thess [59] compared the predictive accuracy of the Boussinesq approximation and

Table 2.5: A summary of recent developments in numerical methodologies for studying PCM melting

Author[ref]	Technique	Key findings
Jiaung et al.[48]	LBM modeling of 1D melting and 2D solidification.	Developed an enthalpy-based LBM (D1Q2, D2Q9); accurately reproduced conduction-driven melting and solidification behavior.
Nedjar [49]	FEM simulation of convection-diffusion and transient heating.	Introduced a temperature-enthalpy formulation. A relaxed linearization method provided flexibility for mushy/isothermal conditions.
Evans et al. [50]	Gallium melting with enthalpy-porosity (FDM).	Coupled SIMPLE with JFNK and BDF time integration; achieved faster convergence and improved stability compared to classical SIMPLE.
Jourabian et al.[51]	Ice melting enhanced by Cu nanoparticles (LBM D2Q9).	Proposed a modified relaxation-time model; improved prediction of melt rate and vortex formation due to increased conductivity.
Jourabian et al.[52]	LBM study of ice melting in heated cavity (D2Q9/D2Q5).	Melt rate is highly sensitive to heater placement. Different positions produced distinct convection patterns and melting durations.
Kaba et al.[53]	Tin melting using stream-function–vorticity.	Developed a modified vorticity-stream-function method. FDM-FVM hybrid discretization produced accurate low-Prandtl melting flow predictions.
El Haddad et al.[54]	CPS-based FEM solution for enthalpy-porosity system.	Used Coupled Projection Scheme with BDF2 integration. Validated accuracy with benchmark data for air, water, and octadecane.
Dutta et al.[55]	Experimental and numerical study in annular cavity.	Combined thermocouple measurements with enthalpy-porosity modeling. Eccentricity and inclination significantly affected melting patterns.

the Volume of Fluid (VoF) method against experimental measurements. Their study shows that the Boussinesq approach generally overestimates velocity magnitudes, although the deviations in temperature and liquid fraction predictions remain small, typically below 4%. This highlights the importance of selecting an appropriate numerical scheme, particularly in convection-driven melting scenarios. The numerical approaches described in this section are summarized in Table 2.5

Based on the collective literature, modelling strategies for PCM heat transfer can be grouped into three major categories: analytical methods, which employ simplified assumptions to obtain closed-form solutions; conduction-based formulations, such as Stefan-type models and effective heat capacity treatments; and convection–diffusion-based formulations, which couple momentum and energy transport to capture natural convection, interface evolution, and turbulence effects. These approaches form the basis of both classical and modern PCM modelling, with early methodologies being continuously refined to incorporate enhanced physical modelling, improved numerical algorithms, and increased computational capability.

The enthalpy–porosity method is widely employed to simulate the coupled convection–diffusion processes occurring in phase-changing materials. By representing the solid–liquid interface as a porous medium, this approach permits the use of a fixed computational grid while solving the momentum and energy equations simultaneously, thereby avoiding the need for coordinate transformations or explicit interface tracking. This method is computationally efficient and has been further enhanced to account for mushy-zone behaviour and the distinct thermo-physical properties of the solid and liquid phases [60].

2.3.1 Lattice Boltzmann Method

The lattice Boltzmann method (LBM) is based on simplified kinetic models derived from the Boltzmann-Maxwell equation, resulting in a mesoscopic framework in which macroscopic fluid behavior emerges from the collective dynamics of microscopic particle populations [61]. In this method, the computational domain is discretized into a regular lattice, with each node representing a limited collection of discrete particle velocity directions that vary as a result of streaming and collision processes. Unlike traditional continuum-based Navier-Stokes solvers, which rely on macroscopic conservation principles, LBM directly models the evolution of particle distribution functions, capturing complex flow physics with simple, local operations.

Furthermore, LBM is still usable across a wider range of Knudsen numbers, covering the continuum and rarefied flow regimes and making it especially well suited for multiscale and microscale flow problems. Its intrinsic parallelism, geometric flexibility, and ease of coupling with phase-change and thermal models make it particularly suitable for complex transport processes.

2.3.2 Stream Function—Vorticity Formulation

The stream function-vorticity formulation is one of the first transient numerical methodologies devised for modeling incompressible flows, particularly in cases when explicit pressure computation is not required. By removing pressure from the governing equations, this approach decreases computing complexity while ensuring mass conservation. In phase-change issues, the primary emphasis is usually in predicting the evolution of the melt front and the accompanying convective flow structures within the liquid region, rather than resolving the pressure field. As melting continues, buoyancy-driven convection becomes increasingly important in heat transmission, necessitating precise depiction of flow patterns.

Several numerical techniques have been proposed to incorporate solid-liquid interface motion into such formulations, including fixed-grid methods, deforming-grid approaches, and front-tracking or front-capturing schemes, each with varying levels of accuracy and computational cost depending on the complexity of the phase-change process [62].

Fixed Grid Approach

The fixed grid technique incorporates the effects of the shifting phase boundary indirectly into the governing equations via new volume source terms, reducing the necessity for explicit interface tracking. This approach is theoretically similar to the enthalpy-porosity technique, which incorporates phase-change phenomena within the formulation itself. The thermophysical state of the phase-changing material is described using an enthalpy-based formulation, with viscosity defined as a function of enthalpy to account for the progressive change in material stiffness during melting or solidification. Furthermore, a porosity-dependent source factor is included in the momentum equations to describe flow resistance in interfacial and mushy regions, effectively reducing velocity in solid zones while enabling fluid mobility in liquid regions. This method produces a smooth and numerically stable depiction of the solid-liquid transition on a set computational grid[63].

Transformed Grid Approach

In the transformed grid or body-fitted coordinate method, the movement of the solid-liquid interface is handled by transforming the spatial and temporal coordinates so that the interface remains stationary in the transformed domain. The transformed governing equations yield a general transport equation defined on a fixed, uniformly spaced computational grid, while the interface movement is preserved in the physical coordinate system. This strategy improves numerical accuracy, especially near curved or evolving boundaries, because it avoids deformation of the computational mesh and ensures more consistent discretisation throughout the domain[63].

Conventional PCM storage systems that utilize horizontal and vertical heat-transfer surfaces often suffer from limited heat-transfer efficiency and slow melting rates, primarily due to the inherently low thermal conductivity of most phase change materials. To overcome these limitations while using the same PCM, performance can be enhanced by increasing the effective heat-transfer surface area. This has led to the development of non-conventional storage designs featuring geometries with expanded surface areas. The influence of these advanced PCM storage configurations on heat-transfer enhancement has been comprehensively reviewed in [74, 75]. Table 2.6 presents work on PCM melting in non-conventional geometries.

The motivation for investigating cyclic heating and cooling arises from the operational characteristics of passive solar energy systems, where energy storage materials are subjected to fluctuating solar radiation. These variations influence the thermal response of phase change materials (PCMs), making it essential to understand their behavior under periodic thermal conditions, as this directly impacts the performance of solar energy applications. Kant et al. [76] analyzed the melting behavior of n-Octadecane under a sinusoidally varying heat flux using COMSOL Multiphysics. Xinyu et al. [77] developed a composite phase change energy storage unit combining water and PCM, incorporating a periodically varying temperature boundary condition. Shahzeb et al. [78] defined four distinct response modes of PCM-based heat sinks and derived predictive expressions for their effectiveness by applying sinusoidal heat input, examining their behavior across a wide range of heat flux amplitudes 100 W/cm^2 and periods (5-640 s). Haocheng et al. [79] explored the thermal management performance of electronic devices equipped with PCMs and pin-fin heat sinks under cyclic heating and cooling. Xinyu et al.[80] conducted numerical simulations to evaluate the impact of a fluctuating heat source on the charging dynamics of the thermal energy storage unit. Table 2.7 summarizes the work done to date on PCM heated with cyclic boundary conditions.

The melting behavior of PCMs is strongly governed by natural convection within the liquid phase, which is influenced by the direction and magnitude of buoyancy forces. As a result, the inclination angle of the enclosure becomes a critical parameter affecting melting dynamics. However, only limited research has explored the role of inclination in PCM melting processes.

Among the numerical studies, Houssam et al. [81] analyzed heat transfer and phase-change characteristics of lauric acid in both rectangular and inclined enclosures at tilt angles of 30°, 45°, and 60°. Their results showed that the rectangular enclosure had the longest melting time and the highest thermal storage capacity, whereas a 45° inclination provided the best practical

Table 2.6: A summary of investigations on melting behavior in various enclosure geometries

Author[ref]	Configuration	Key findings
Mishra et al.[64]	Circular, Semicircular, Square, Rectangular, Ellipsoidal & Triangular.	The inverted semi-circular enclosure exhibits the most effective melting behavior relative to the other shapes analyzed.
Sodmand et al.[65]	Rectangular, Triangular & cylindrical enclosures.	The cylindrical configuration consistently exhibits a greater liquid fraction during solidification than other geometries.
F.L. Tan et al.[66]	Spherical enclosure.	Conduction governs the early phase; however, buoyancy-induced flow becomes significant as the liquid fraction grows.
Dhaidan et al.[67]	Right angled triangular enclosures.	Increasing hot-wall temperature and reducing storage unit dimensions results in a faster melting rate.
Haddad et al.[68]	Isosceles triangular cavity.	At the end of melting, the Lorentz force has a limited effect with a horizontal magnetic field and no effect with a vertical one.
Haddad et al.[69]	Trapezoidal enclosure with waviness.	The study shows that the convection cell count depends on both time and wave conditions.
Tabassum et al.[70]	Equilateral triangular enclosure.	Energy storage is strongly influenced by the configuration and whether the inner tube is concentric or eccentric.
Tabassum et al.[71]	Equilateral triangular enclosure.	Melting rate of impure PCM is strongly affected by inner tube wall temperature and initial PCM temperature.
Iachachene et al.[72]	Trapezoidal enclosure.	A decline in heat transfer is observed in NEPCMs when thermal conductivity enhancement is less than 80%.
Sharma et al.[73]	Trapezoidal enclosure.	The total solidification time decreases for both pure fluid and nanofluid at all nanoparticle concentrations.

performance by balancing energy storage and melting duration. Tushar et al. [82] investigated PCM melting in an inclined finned enclosure heated from the top and reported that melting time increased with lower inclination angles and with the use of extended surfaces.

Table 2.7: A summary of Periodic/cyclic boundary condition investigations on PCM melting

Author[ref]	Configuration	Key findings
Kant et al.[76]	$q'' = 250 \cdot \pi \cdot \sin\left(\frac{\pi \cdot t}{45}\right)$ where $(0 < t < 90)$	Melting time decreases with variable heat flux compared to constant heat flux. Melting process occurs faster than solidification for the PCM.
Xinyu et al.[77]	$T = T_w + A \cdot \sin(\pi \cdot t/\tau)$	Amplitude and time period influence sensible heat release, while latent heat remains largely unaffected due to combined heat sink and convection effects.
Shahzeb et al.[78]	$q'' = q''_0 + q''_A \cdot \sin(\omega \cdot t)$	PCMs with wider melting ranges perform effectively under a broader spectrum of heating conditions, a design insight with limited previous attention.
Haocheng et al.[79]	The heat source is a sinusoidal power load (0–30 W peak, 3.1 min heating + 6.3 min cooling cycles)	Compared to conventional PCM, the MEPCM offers significant advantages in thermal management.
Xinyu et al.[80]	$T = T_w + A \cdot \sin(2\pi \cdot t/\tau)$	Increasing amplitude and initial temperature rises average energy storage rate. Extending the half-period causes the rate to initially increase then decrease.

Dominic et al. [83] studied PCM melting in rectangular enclosures with and without fins for photovoltaic (PV) panel temperature control and found that optimal thermal management is achieved using full-width fins attached to both front and back plates. Mohit et al. [84] examined RT42 PCM melting in a 2D LTES system equipped with rectangular, triangular, and Y-shaped fins under upright ($\theta = 90°$) and inclined ($\theta = 35°$) configurations, reporting significant Nusselt number enhancements up to 2.67 times for upright and 2.32 times for the inclined Y-fin configuration. Sheng et al. [85] analysed paraffin-copper foam composite PCM melting at multiple tilt angles.

From the experimental perspective, Abdel et al. [86] studied transient melting of paraffin in a square enclosure at 0°,45°, and 90°, tracking the solid–liquid interface using thermocouples and infrared imaging. The findings demonstrated that inclination strongly alters natural convection and melting progression, reducing total melting time by approximately 56% and 48% for bottom-heated and inclined cases, respectively, compared with the vertical arrangement. R. Kothari et al. [87] experimentally evaluated inclination effects on PCM-based heat sinks with and without fins at heat fluxes of 1.3, 2 and 2.7 kW/m^2 and tilt angles from 0° to 90°. The study found that inclination had little impact on non-PCM configurations, whereas for PCM-based systems, reducing inclination from 90° to 0° increased operating time by 66%, 45%, and 43% for systems with zero, one, and two fins, respectively, at a heat flux of 2.0 kW/m^2. Table **??** summarises the experimental and numerical findings on PCM melting with orientations.

Mehmet et al. [88] examined RT-50 melting in a cylindrical enclosure with a central microchannel and demonstrated that inclination enhanced natural convection, accelerating melting and heat transfer. Wuyang Li et al. [89] investigated a PCM-integrated solar chimney and showed that inclination affects buoyancy-driven convection and melting duration, with melting time increasing by 6.7% and 8.6% at 45° and 30°, respectively, relative to 60°. Furthermore, melting duration increased by 16.1% and 8.0% at heat fluxes of 400 and 500 W/m^2 compared to 600 W/m^2, indicating that inclination had a stronger influence than the applied heat flux on heat transfer performance.

2.4 Existing gaps in research

- Detailed depiction of Rayleigh–Bénard convection in semi-circular enclosures is not widely researched
- Rayleigh–Bénard convection under pulsed heat inputs with time scale similar to that of convection is not available in literature
- Rayleigh–Bénard convection under tilted orientation of confinements under heating using patterned surfaces is not available
- Detailed analysis of melt dynamics and heat transfer correlations are not available in above cases

2.5 Research objectives

The main research objectives of the present study are

- To develop an experimental set-up to measure temperature response, visualize the melt front development and convection pattern in PCM.
- To develop a computational procedure based on enthalpy-porosity method finite volume method with provisions for applying time-varying heat flux boundary conditios for the simulation of melt dynamics.
- To analyze thermal convection physics in phase changing material in semi-circular confinement under constant heat flux conditions.
- To analyze thermal convection physics in phase changing material subjected to heat flux fluctuations at the heat transfer boundary.

3 Methodology

This chapter presents the methodology adopted for both the numerical and experimental investigations. The Shadowgraph technique is utilized to visualize the flow field, while temperature measurements recorded using thermocouples are employed to validate the numerical model. The numerical approach detailed in this chapter includes solving the Navier–Stokes equations, applying no-slip boundary conditions, using the enthalpy-porosity method, and incorporating temperature-dependent thermophysical properties. The chapter concludes with a description of the validation procedure based on experimental data, along with the assessment of grid and time step independence.

Introduction

Experimental and numerical investigations have been carried out on phase change materials (PCMs) under various thermal boundary conditions and enclosure configurations in this work. During the melting process, the dominant heat transfer mechanism transitions from conduction to natural convection as the liquid fraction increases, depending on the temperature gradient and material properties. Numerical modeling using the Navier–Stokes equations combined with the enthalpy-porosity method can provide detailed insight into the thermal behavior and phase transition dynamics with relatively lower experimental effort.

3.1 Experimental Methodology

The schematic of a typical experimental setup used for study is shown in Fig 3.1 The PCM storage unit consists of a semicircular chamber Fig 3.2 of radius 60 mm . The front and rear walls are constructed from 5 mm thick acrylic sheets. To accommodate thermal expansion of the PCM, the chamber is open at the top and exposed to ambient conditions. The acrylic sidewalls are mounted vertically on a well-insulated aluminum heater casing with rectangular slots. A 350 W cartridge heater (Make: Ideal Heaters), 10 mm in diameter and 120 mm in length, is firmly embedded within the casing. Thermal grease is applied to the heater surface to ensure efficient heat transfer to the casing, while a high-temperature sealant (Anabond) is used to prevent leakage of molten PCM. A regulated power supply provides a controlled and uniform surface heat flux to the bottom of the PCM chamber. To capture the thermal response at different positions, twelve K-type thermocouples are installed as shown in Fig 3.3. The thermocouples are placed in sequence to minimize interference from convection currents. Each trial is conducted with identical heat input conditions. Before testing, the thermocouples are calibrated by immersion in boiling water, using a high-precision thermometer as a reference. The average uncertainty in thermocouple measurements is 0.084 °C, corresponding to an error margin of 2.78%. Temperature data are recorded at a frequency of 3 Hz using a data acquisition system. To achieve the desired inclinations, an aluminum angle fixture is used, which allows flexible adjustment and can be fixed at any required angle. Figure 3.4 and 3.5 shows the shadowgraphy arrangement and heat transfer surfaces considered in this study. The first configuration is a smooth flat surface with a heat transfer length of 144 mm. The second employs a square patterned surface, with consecutive square elements spaced 1.935 mm apart, each having a width and depth of 2 mm. The third is a triangular patterned surface, where consecutive isosceles triangular elements, each 2 mm in height, are arranged in sequence. In both roughened configurations, the number of roughness elements is limited to 30. Since the cartridge heaters utilized in the present study are designed for alternating current (AC) operation, AC heating was adopted. These heaters are found to offer controlled heating suitable for the experiments envisaged in the current study.

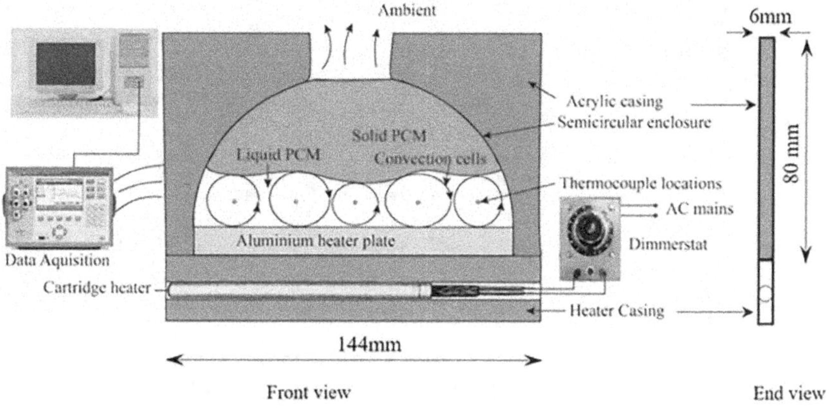

Figure 3.1: Experimental setup details semicircular enclosure

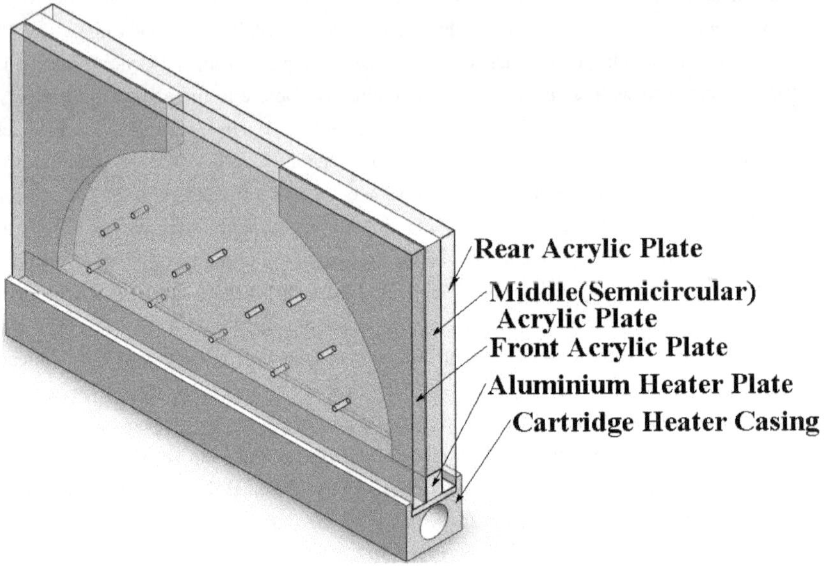

Figure 3.2: Isometric view of the semicircular enclosure

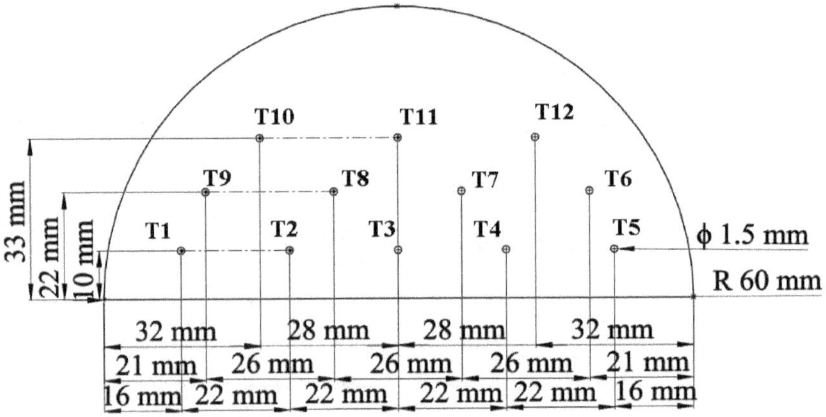

Figure 3.3: Thermocouple locations

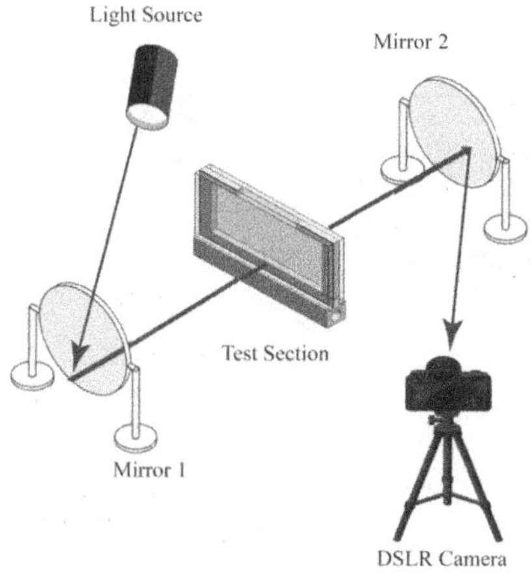

Figure 3.4: Z-type shadowgraph arrangement

A Z-type shadowgraph configuration, employing two concave mirrors each with a focal length of 150 cm, is used to visualize refractive-index gradients during PCM melting. A collimated light source is directed onto the first concave mirror, reflected through the test enclosure and then onto the second mirror which focused the perturbed beam onto the imaging sensor. The long focal length increased sensitivity to small density/refractive-index changes while keeping the imaging geometry compact. Mirrors are mounted on adjustable kinematic mounts for precise alignment as shown in Fig 3.4 neutral density filtering and camera exposure control are used to avoid saturation. The wavy surface is defined by a sinusoidal function,

$$y = sin(0.3x) \tag{3.1}$$

The corrugated surface is formed by a sequence of connected arc segments, each with a radius of 10 mm. In all configurations, the heat transfer surface width is maintained at 6 mm. The experimental procedure described here is also reported in chapter 6, where the heating surfaces are square patterned and triangular patterned elements surfaces, and the enclosure is rectangular.

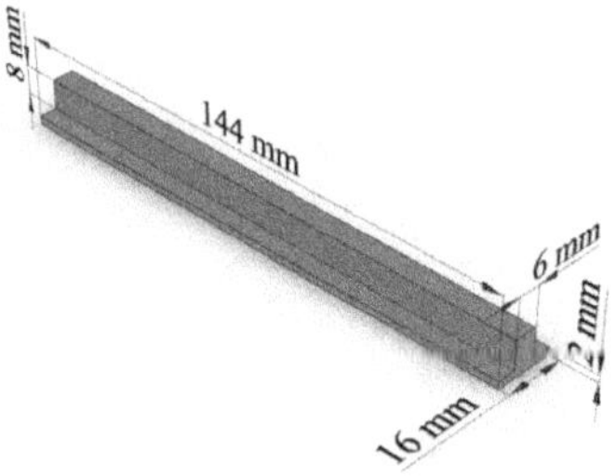

(a) Flat surface

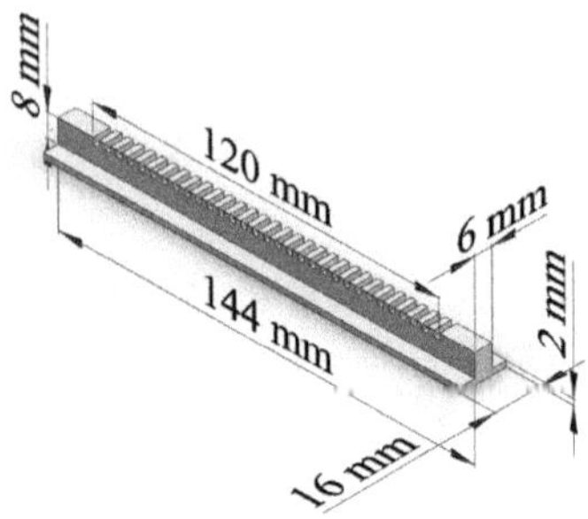

(b) Surface with square patterned elements

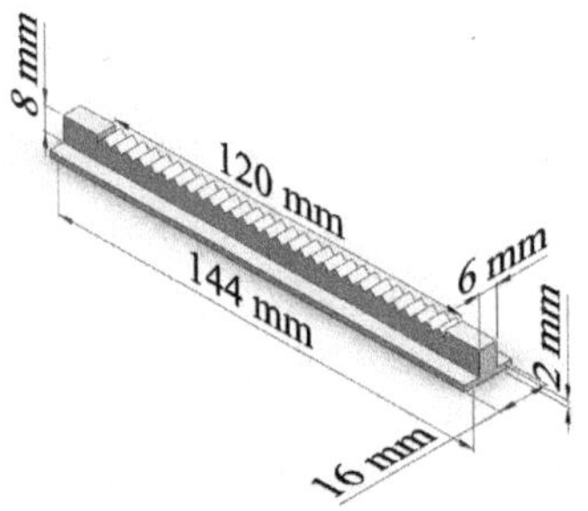

(c) Surface with triangular patterned elements

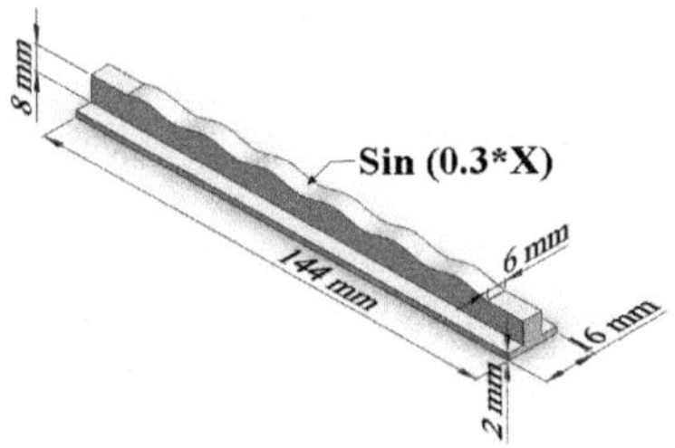

(d) Wavy Surface

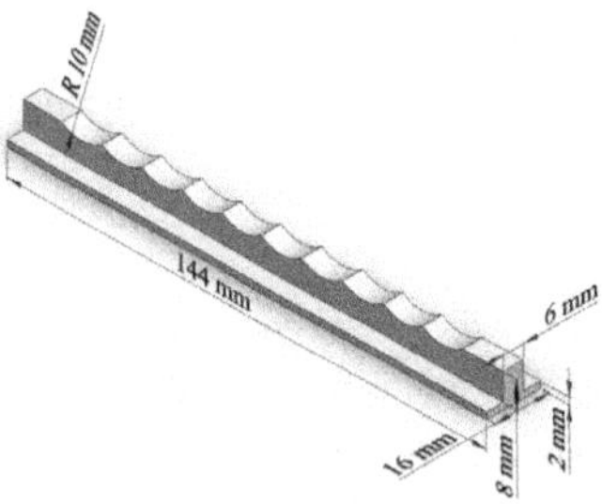

(e) Arc-Corrugated Surface

Figure 3.5: Different heat transfer surfaces employed in this study

3.2 Numerical Methodology

The thermal management system employing phase change material (PCM) is represented using a two-dimensional numerical model, as illustrated in Figs 3.6 , 3.7 and 3.8. The geometries considered in the present study, corresponding to the experimental configurations used for validation. The melted (liquid) PCM within the enclosure is assumed to behave as a Newtonian and incompressible fluid, and the flow in the liquid region is considered laminar. The effects of viscous dissipation are neglected. The thermophysical properties of the PCM are dependent on temperature and phase . Based on the above assumptions, the governing equations describing the melting process of lauric acid are expressed as follows Mass conservation

$$\frac{\partial \rho}{\partial t} + \nabla.(\rho u_i) = 0 \tag{3.2}$$

Momentum conservation

$$\frac{\partial \rho u_i}{\partial t} + \nabla.(\rho u_i u_j) = -\frac{\partial p}{\partial x_i} + \nabla.(\mu \nabla(u_i) + \rho \beta \vec{g}(T\text{-}T_{ref}) + S_i \tag{3.3}$$

Energy equation

$$\frac{\partial \rho H}{\partial t} + \nabla.(\rho u_i H) = \nabla.(k \nabla T) \tag{3.4}$$

The dimensionless temperature, θ, is defined following the approach of Zongqin [90]. It is not incorporated into the governing equations; rather, it is used for plotting dimensionless temperature contours superimposed with streamlines to facilitate the visualization of thermal and flow fields in the present study.

$$\theta = \frac{T\text{-}T_{ref}}{\Delta T_{max}}; \; \Delta T_{max} = T_{avg}\text{-}T_{ref} \tag{3.5}$$

Here, T_{avg} is the length averaged temperature of heating surface

$$T_{avg} = \frac{1}{L}\int_0^L T_w(y)\, dy \tag{3.6}$$

and T_{ref} is the ambient temperature. In this model, the mushy region is represented using a porous medium approximation. The source term, enthalpy, and liquid fraction in the governing equations are expressed as:

$$S_i = A_m \frac{(1\text{-}L_F^2)}{(L_F)^3 + \epsilon} \tag{3.7}$$

$$H = h + \Delta H \tag{3.8}$$

$$h = h_{ref} + \int_{T_ref} C_P\, dT \tag{3.9}$$

$$L_F = \frac{\Delta H}{F_l} = 0 \text{ if } \quad \mathrm{T} \leq T_S;\ 1 \text{ if } \mathrm{T} \geq T_l; \frac{(T\text{-}T_S)}{T_l\text{-}T} \text{ if } T_S < T < T_l. \tag{3.10}$$

To investigate the physics of this problem, several key non-dimensional numbers are employed. Zongqin [90] formulated the Rayleigh number (Ra) to incorporate the effects of melt interface dynamics (h) and heat flux q" across the heated surface. Rayleigh number

$$Ra = \frac{g\beta q"(h_m)^4}{k\alpha\nu} \tag{3.11}$$

The non-dimensional timescale (F_O) is employed to examine the melt evolution under different heat transfer surface configurations.

$$F_O = \frac{\alpha\tau}{W^2} \tag{3.12}$$

Furthermore, the heat transfer characteristics are evaluated using the evolution of the Nusselt number under constant heat flux conditions [90], defined as:

$$Nu = \frac{q"(h_m)}{k(T_{avg}\text{-}T_m)} \tag{3.13}$$

Similarly, the variation of the thermophysical properties of lauric acid is modeled following the approach of Mukhesh et al. [40]. The general relations for these property variations are provided below:

$$Density(kg/m^3) = \frac{885}{0.001(T\text{-}316.35) + 1} \tag{3.14}$$

$$DynamicViscosity(kg/ms) = 0.0022exp(\text{-}4.25 + \frac{1790}{T}) \tag{3.15}$$

Thermal conductivity (W/mK)

$$= 0.16 \text{ if } \mathrm{T} < T_s$$

$$= 0.16\cdot(1\text{-}f) + 0.14\cdot f$$

$$= 0.14 \text{ if } T > T_l$$

Heat capacity (J/kgK) = 2285
Latent Heat of Fusion, (J/kg) = 187210

The numerical simulation of PCM melting in an enclosure with a wavy bottom-heated surface is conducted using the finite volume method based on the Semi-Implicit Method for Pressure Linked Equations-Consistent (SIMPLEC) algorithm, as implemented in the commercial software ANSYS FLUENT. The enthalpy-porosity technique is employed to model the phase transition during melting.

Figure 3.6 presents the semicircular enclosure with a radius of 60 mm, where a constant heat input is applied at the bottom surface. The curved side walls are treated as adiabatic, while the upper 40 mm region is exposed to the ambient environment at 303 K with a convective heat transfer coefficient of $h = 10W/m^2K$. To investigate the influence of heater surface geometry on PCM melting behavior, three surface configurations are considered, as shown in Figure 3.5: (a) a flat surface, (b) a wavy surface, and (c) an arc-corrugated surface. A corresponding two-dimensional representation of these heating surfaces is adopted in the numerical model. The heat input values considered in this study are 5.04 W, 5.76 W, 6.48 W, and 7.2 W.

Figure 3.7 illustrates the computational domain, consisting of a square enclosure with a 25 mm side length and a wavy heated surface at the bottom. The top boundary and both vertical walls are treated as adiabatic. The melting process within this confined geometry is considered two-dimensional, with flow and thermal convection occurring in the x–y plane.

Figure 3.8 illustrates the computational domain of the inclined rectangular enclosure (120 mm length and 50 mm height). The side walls are treated as insulated, while the bottom surface is subjected to a uniform heat flux. The top boundary is exposed to the ambient environment at 303 K with a convective heat transfer coefficient of $h = 10W/m^2K$. The influence of surface texture on the melting behavior is evaluated using surfaces with square and triangular patterned elements, as shown in Figure 3.5 b and c, respectively. It should be noted that Figures 3.6, 3.7, and 3.8 correspond to three separate computational domains used for different investigations in the present study. Figure 3.6 presents the computational domain of a semicircular enclosure with a radius of 60 mm heated from the bottom under a constant heat flux. Figure 3.7 shows a square enclosure of side length 25 mm incorporating a wavy heating surface at the bottom, while the side walls are maintained adiabatic. Figure 3.8 illustrates a rectangular enclosure measuring 120 mm × 50 mm. Each of these computational domains is employed for a different study and is therefore presented separately in chapter no. 4, 5 and 6 respectively.

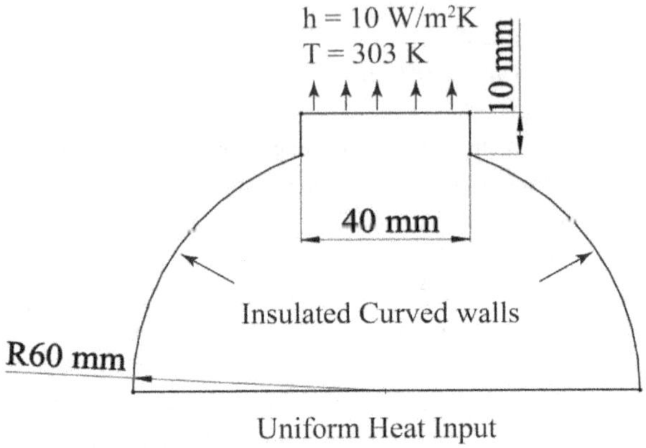

Figure 3.6: Computational domain of semicircular enclosure under constant heat input

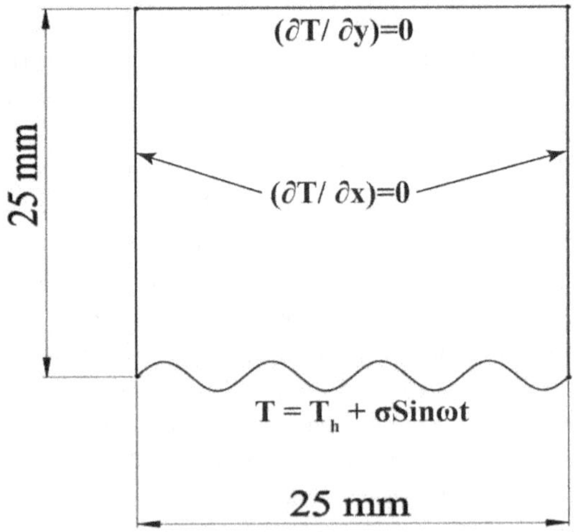

Figure 3.7: Computational domain of wavy surface with periodic boundary condition

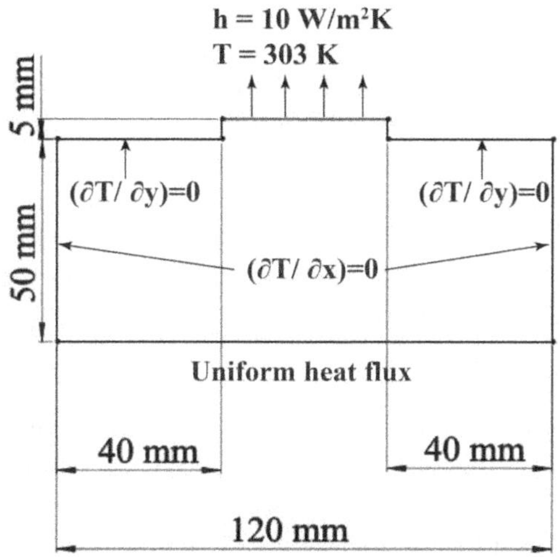

Figure 3.8: Computational domain of rectangular enclosure with inclination

3.3 Validations and grid independence/time-step independence studies

The simulation outcomes exhibit strong agreement with the experimental data, confirming the reliability of the numerical methodology adopted in this study. In both numerical and experimental analyses, the progression of the melt fraction demonstrates a linear relationship with the Fourier number. A grid independence study is performed by comparing melt fraction behavior across three mesh densities, with the coarsest grid comprising 15,000 nodes and the finest containing 60,000 nodes. Figure 3.10 illustrates the melt fraction evolution for these grid resolutions. The coarsest mesh predicts a slightly higher melt fraction, while the medium and fine grids yield nearly identical results. Considering the balance between computational cost and accuracy, the grid with 30,000 nodes is selected for subsequent simulations.

Beyond melt fraction analysis, model validation is further evaluated by comparing the non-dimensional temperature (θ) evolution with respect to non-dimensional time (F_O). Figure 3.11 presents the comparison of simulation predictions and experimental measurements at predefined thermocouple locations. A strong correlation is observed across all cases. Initially, temperature rises slowly due to diffusion-dominated heat transfer. The onset of flow instability and the development of convection currents accelerate the temperature increase, with these convective effects also causing observable oscillations in the temperature profiles. The temperature validation figure 3.11 and the melt fraction validation figure 3.12 have maximum percentage deviations of 3.36 and 2.58, respectively.The other heating surface configurations considered in the study, namely the wavy and corrugated surfaces, exhibit similar trends under the same heat input, with only minor quantitative differences. A qualitative comparison between the numerical and experimental shadowgraphy images of convection cells formed in the liquid region over wavy and corrugated heating surfaces is presented in Fig. 3.13. The numerical shadowgraph images are generated using the normalized density-gradient magnitude. In both configurations, plume structures emerging above the curved surfaces are distinct, with the wavy surface producing a curved melt front aligned with its geometry, whereas the corrugated surface results in an almost straight melt front. The convection cells also introduce oscillations in the temperature field, which are captured in the quantitative comparison of the non-dimensional temperature.The domain is discretized using quadrilateral finite control volumes, with the discretization pattern near the heating surface illustrated in Fig. 3.14.

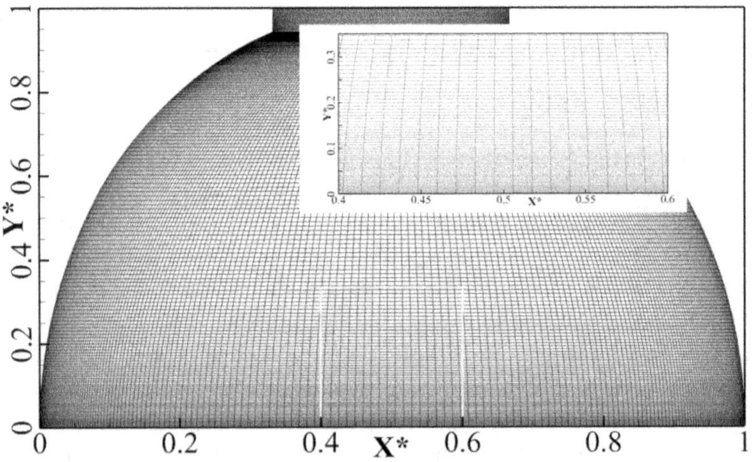

Figure 3.9: Mesh details of semicircular enclosure

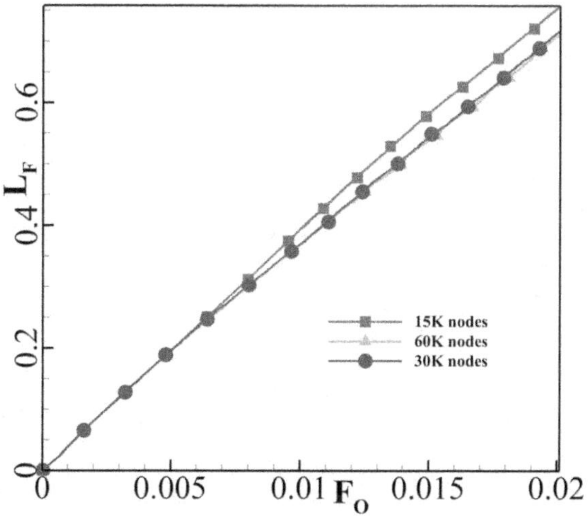

Figure 3.10: Grid-independence study (semicircular enclosure)

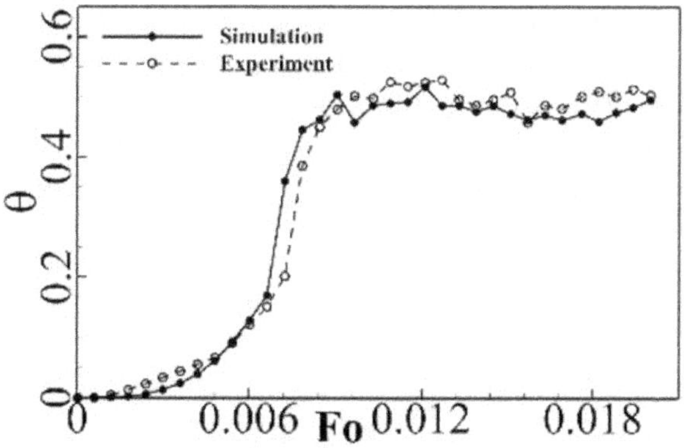

Figure 3.11: Non-dimensional experimental temperature validation (semicircular enclosure)

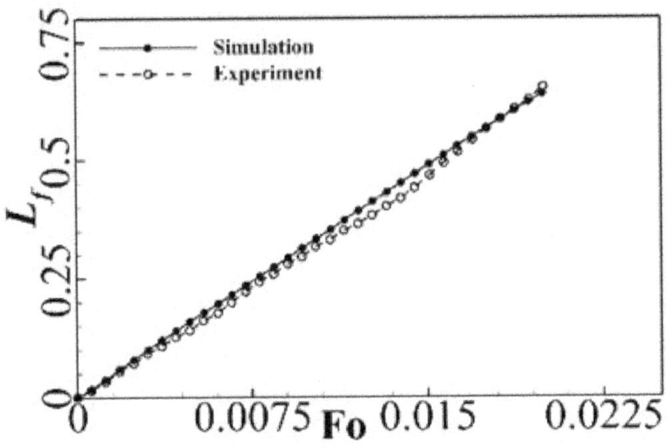

Figure 3.12: Experimental liquid fraction validation (semicircular enclosure)

A grid independence test is carried out for the case of AWR = 1 and $\sigma = 0$ K by progressively refining the mesh. Unsteady simulations are performed for a total duration of 700 s, and the time-averaged Nusselt number obtained for different grid refinements is compared. In addition, a detailed real-time comparison of the predicted liquid fraction across various grid sizes is presented in Fig. 3.15 Based on these assessments, a grid size of 150 x 150 is selected for the subsequent computational studies.A time-step independence test is also performed by successively reducing the Courant–Friedrichs–Lewy (CFL) number to 0.02, 0.01, and 0.005. The predicted melting time for the last two CFL values differs by less than 1.2%, confirming time-step independence. Furthermore, the adopted convergence criteria ensure that, for the chosen time-step size, all conservation equations reach a residual level below 10^{-6}. at every time step.

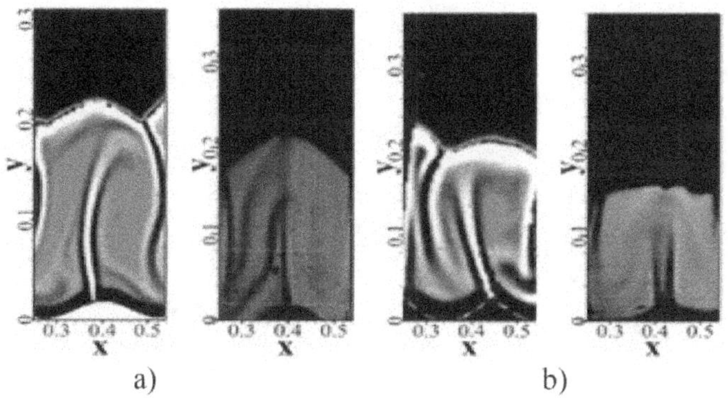

Figure 3.13: Comparison of convection cells formed over the (a) wavy and (b) corrugated heat transfer surfaces, as observed in numerical simulations and experimental shadowgraphy, respectively.

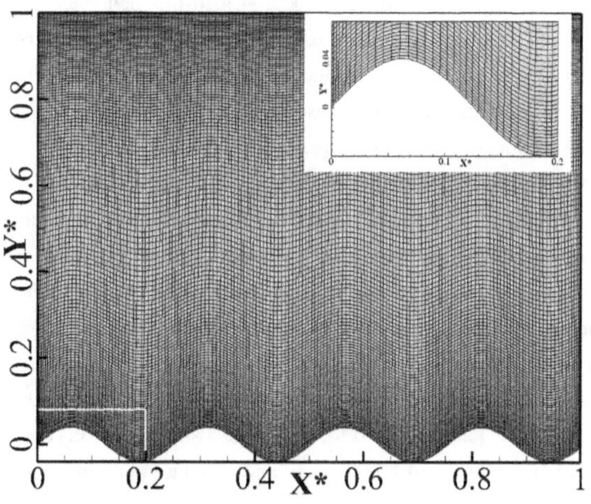

Figure 3.14: Mesh details of square enclosure with wavy surface

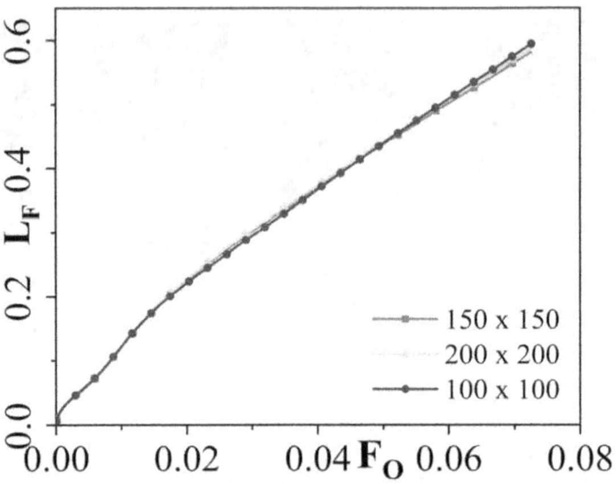

Figure 3.15: Grid-independence study of square enclosure with wavy surface

The computational domain is discretized using a structured quadrilateral mapped mesh. To accurately capture the steep gradients arising from conjugate heat transfer, the mesh is appropriately refined near the heating surface. A grid-independence study is carried out using three mesh resolutions, comprising approximately 40,000, 80,000 and 160,000 elements. Figure 3.17 presents a comparison of the computed liquid fraction across these grid levels. As the results from the second and third mesh resolutions show close agreement, the second-level grid (approximately 80,000 elements) is adopted for further simulations.

Additionally, a time-step independence analysis is performed using this grid, with computations conducted at CFL numbers of 0.02, 0.01, and 0.005. The variation in total melting time between CFL = 0.01 and CFL = 0.005 remains within a 0.546% margin. Consequently, all numerical simulations in the present study are carried out using the second-level grid and a CFL number of 0.01. Figure 3.16 shows the selected computational grid used in the present study. In all figures, the Cartesian coordinates (x and y) are non-dimensionalized, with the heating surface length taken as the characteristic horizontal length and the enclosure height as the vertical reference. The grid used for the numerical simulations corresponds to the second refinement level and consists of 80,000 regular quadrilateral elements, as illustrated.

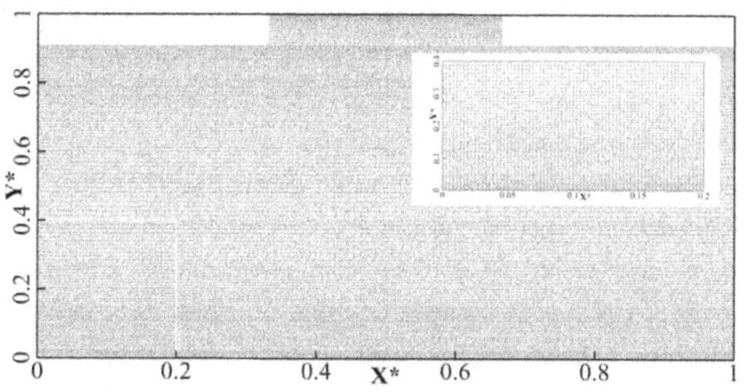

Figure 3.16: Mesh details of rectangular enclosure

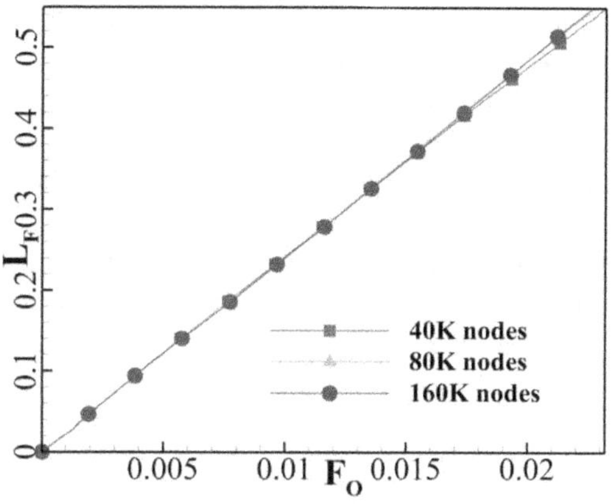

Figure 3.17: Grid-independence study (rectangular enclosure)

Given that heat transfer through a wall is a key aspect of the problem, accurately capturing the boundary layer development near the heated wall is essential. Therefore, a refined mesh is employed near the bottom wall to resolve the boundary layer evolution under constant heat flux conditions. The inset views in Fig. 3.19 highlight the mesh refinement applied near the heating surface for both passage geometries. The computational model is validated through the comparison of melt fraction obtained from simulation results with the results of experimental image analysis. In Figure 3.18 presents a comparison of the melt fraction evolution obtained from both the experiments and numerical simulations for two cases with different heat transfer rates. The melt fraction in the experimental snapshots is determined using image processing in MATLAB, where it is calculated as the ratio of the melted region area to the total PCM area. Further validation is performed by comparing the evolution of dimensionless temperature θ with respect to the Fourier number (F_O) at selected thermocouple locations. As shown in Figure 3.19, the simulated and experimental temperature profiles exhibit strong agreement across all measurement points. During the early diffusion-dominated stage, temperature rises gradually, whereas the onset of thermal instabilities triggers natural convection, leading to a steeper temperature increase. The observed oscillations are attributed to convective mixing effects. The absolute maximum percentage deviation in the experimental melt fraction validation (Figure 3.18)and temperature validation (Fig 3.19) is 2.86 and 1.12, respectively. Figure 3.20 shows the validation of the numerical shadowgraph image against the experimental shadowgraph image for the flat surface case at a heat flux of 7 kW/m^2 and Fo = 0.009773 at 0° inclination. Both images exhibit strong agreement, as the number of convection cells ($\approx$ 8) and their structures are nearly identical. Fig 3.20(b) presents the validation for the case of a surface with triangular patterned elements at a 50° inclination. In this case, the solid–liquid PCM interface serves as the basis for validation, and the evolution of the interface shows close similarity between the experimental and numerical shadowgraph images. Fig 3.20(c) Validation of the experimental shadowgraph image for the surface with square patterned elements at 0° inclination and a heat flux of 8 kW/m^2 (Fo = 0.0058) with the corresponding numerical shadowgraph. Similar to case Fig 3.20 (a), the comparison focuses on the number and structure of convection cells, both of which show strong agreement.

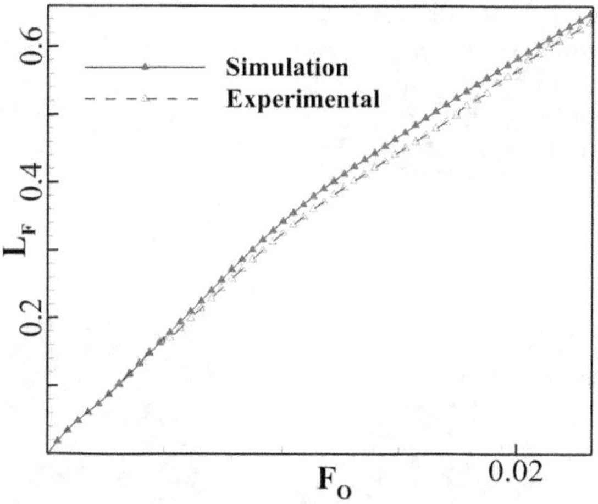

Figure 3.18: Experimental liquid fraction validation (rectangular enclosure)

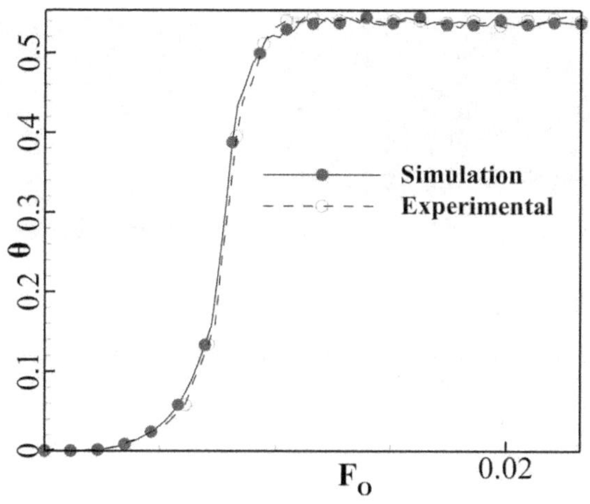

Figure 3.19: Experimental Non-dimensional temperature validation (rectangular enclosure)

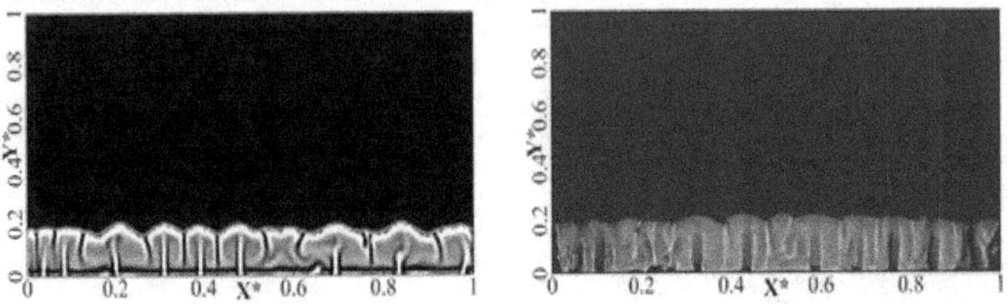

(a) Flat surface 0° inclination, 7 kW/m^2, $F_O = 0.009773$

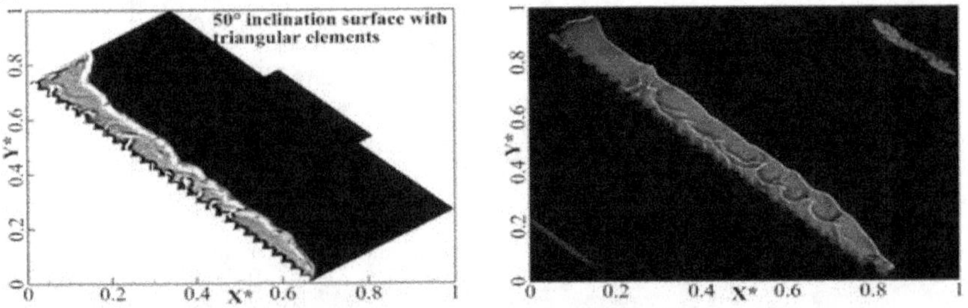

(b) Surface with triangular patterned elements 50° inclination, 8 kW/m^2, $F_O = 0.0087$

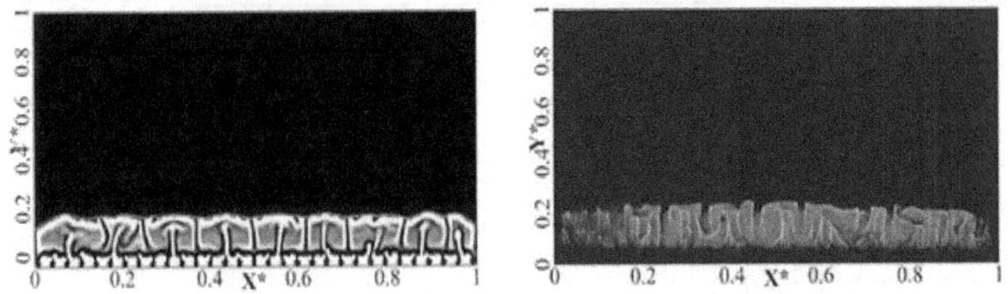

(c) Surface with square patterned elements 0° inclination, 8 kW/m^2, $F_O = 0.0058$

Figure 3.20: Comparison of experimental shadowgraph images with Numerical shadowgraph,left column Numerical shadowgraph images, right column experimental shadowgraph images.

4 Effect of heater surface geometry in melting dynamics and heat transfer

This chapter presents a detailed investigation of melting dynamics and heat transfer over wavy, flat and corrugated surfaces for latent heat energy storage applications. A comprehensive experimental as well as numerical analysis of different surface configurations operating under varying heat inputs is carried out. The flow characteristics within the 2-D semicircular enclosure are examined, including the number of plumes and the associated Y-velocity contours. Additionally, empirical Nu-Ra and δ_T-Ra correlations are proposed to predict the influence of surface geometry on heat-transfer performance.

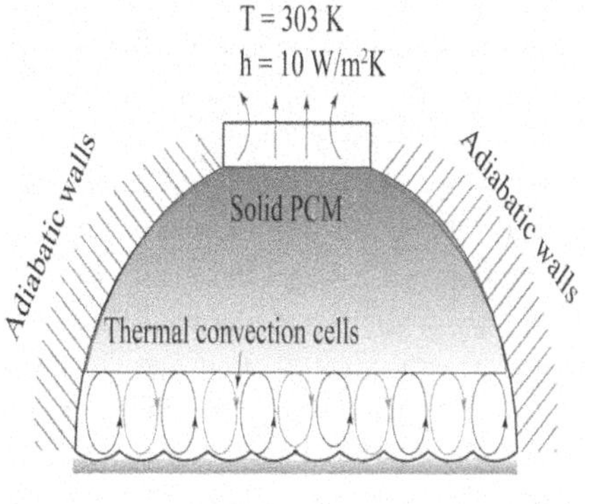

Figure 4.1: Schematic of melting inside semicircular enclosure under corrugated heating surface

Figure 4.1 depicts a two-dimensional semicircular enclosure (radius 60 mm) with a corrugated heating surface. The curved walls are applied under adiabatic conditions. The heating

A detailed presentation of the results presented in this chapter has been published as
Osman, M.F., Sarath, K.P. and Deepu, M., 2024. Studies on the melting dynamics of PCM in a semicircular cavity with straight and wavy heating surfaces. International Communications in Heat and Mass Transfer, 151, p.107249.
doi : `https://doi.org/10.1016/j.icheatmasstransfer.2024.107249`

surface has a uniform heat input boundary condition, and the top surface is open to the environment. In this chapter, three different heating surfaces are used: flat, wavy, and corrugated. The heat input over the surface ranges from 5.04 W to 7.2 W, hence the flat surface will have the largest heat flux for the same heat input in all three heating surfaces (because of smallest area).

4.1 Meltfront evolution and thermal convection physics

At $F_O = 0.016$, notable differences in the flow structures emerge between the three bottom-surface geometries. The flat surface experiences the highest heat flux, generating the strongest buoyant forces in the melted PCM. This enhanced buoyancy drives vigorous upward motion, which accelerates melting and leads to larger oscillations of the melt front. In comparison, the wavy and arc-corrugated surfaces receive slightly lower heat fluxes, resulting in weaker convective driving. Because the flat surface heats the fluid uniformly across its entire length, the rising warm liquid forms a strong and coherent upward flow that effectively pushes the interface upward and produces visible fluctuations. For the wavy surface, buoyant motion originates from many small regions with varying local orientations, causing the vertical flow to split into several smaller convection paths. The arc-corrugated geometry further smooths these effects, guiding the flow into broader but weaker upward streams that have less influence on the interface. Moreover, the flat surface fully exposes the liquid to uniform heating, allowing heat to enter the bulk fluid more efficiently. In contrast, the troughs and arc pockets on the wavy and corrugated surfaces can trap heat near the wall, slowing its transfer into the melt and reducing the strength of vertical motion. Consequently, the flat surface promotes strong, organized convection that mixes and expands the melt region rapidly, whereas the wavy and arc surfaces generate fragmented flow that limits heat transport, leading to slower melting and smaller interface oscillations.

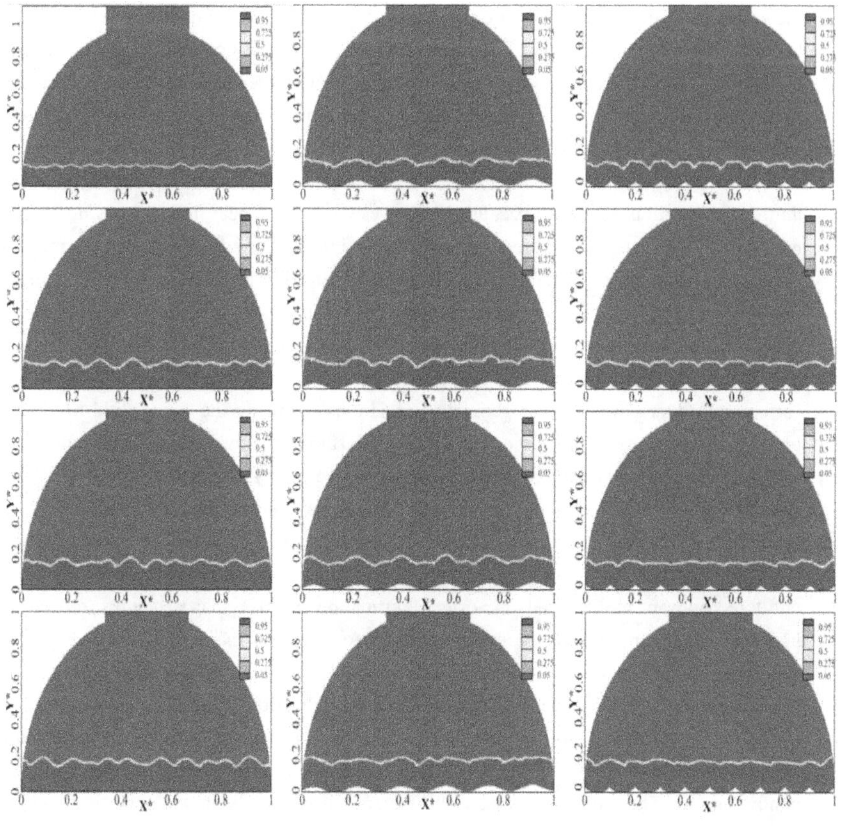

(a) (b) (c)

Figure 4.2: Evolution of the melt fraction within the semicircular enclosure featuring (a) straight, (b) wavy, and (c) corrugated heat-transfer surfaces at a Fourier number of 0.005

It can be observed from figure 4.2, the melting process evolves through three distinct regimes. Initially, a conduction-dominated regime is observed in which the melt front remains almost parallel to the heated surface; this stage is short-lived and typically persists only up to Fo $\approx$ 0.001 for all cases. As soon as temperature gradients intensify near the bottom wall, the system transitions into a linear convection-onset regime, where the steep thermal gradient between the heating surface and the interface generates several small buoyant plumes, leading to the formation of multiple fine-scale vortices. With further melting, these vortices begin to interact and merge, marking the onset of the coarsening regime, during which the initially small convection cells reorganize into fewer, larger, and more stable vortical structures that dominate the flow field. As melting advances, the flat surface generates the strongest and most uniform upward buoyant flow because it receives the highest heat flux and has no geometric modulation. This strong, coherent upward motion interacts continuously with the melt front, causing multiple large oscillations and eventually distorting the vortex structures within the liquid. The wavy surface, however, breaks the upward motion into several smaller convection paths, but one central upflow becomes dominant, producing a single major oscillation at the interface while the rest of the interface remains comparatively smoother. The arc-corrugated surface introduces even gentler geometric variations, allowing the rising fluid to align into broad, uniform convection cells; as a result, the interface remains nearly flat and stable.

In terms of flow structures, the arc-corrugated geometry encourages vortices of similar size and shape because the surface curvature distributes buoyancy more evenly across each arc(Figure 4.3). The wavy surface, with its sharper and more frequent slope changes, generates more vortices that vary in strength and spacing. In the flat surface case, the intense and unfiltered buoyant motion leads to vortex deformation and irregular structures, reflecting the highly energetic and unstable convection field. As a whole, increasing surface smoothness progressively reduces oscillations at the interface and stabilizes the vortex field.

Figures 4.4 and 4.5 present the experimental visualizations of melt fraction, showing the progression of the melt front at different Fourier numbers, along with shadowgraph images capturing gradient variations in the molten flow field for various Fourier numbers and surface configurations. These observations are subsequently used to validate the numerical simulation results.

The corrugated surfaces modify the thermal boundary layer development by creating preferential locations for thermal plume generation. For the wavy and arc-corrugated surfaces, the crests and apex regions act as plume initiation sites due to the reduced thermal resistance and stronger local buoyancy forces. The interaction of these localized thermal plumes with the surrounding PCM enhances mixing and alters the boundary layer structure, resulting in a slightly weaker dependence of boundary layer thickness on Rayleigh number compared with the flat surface.

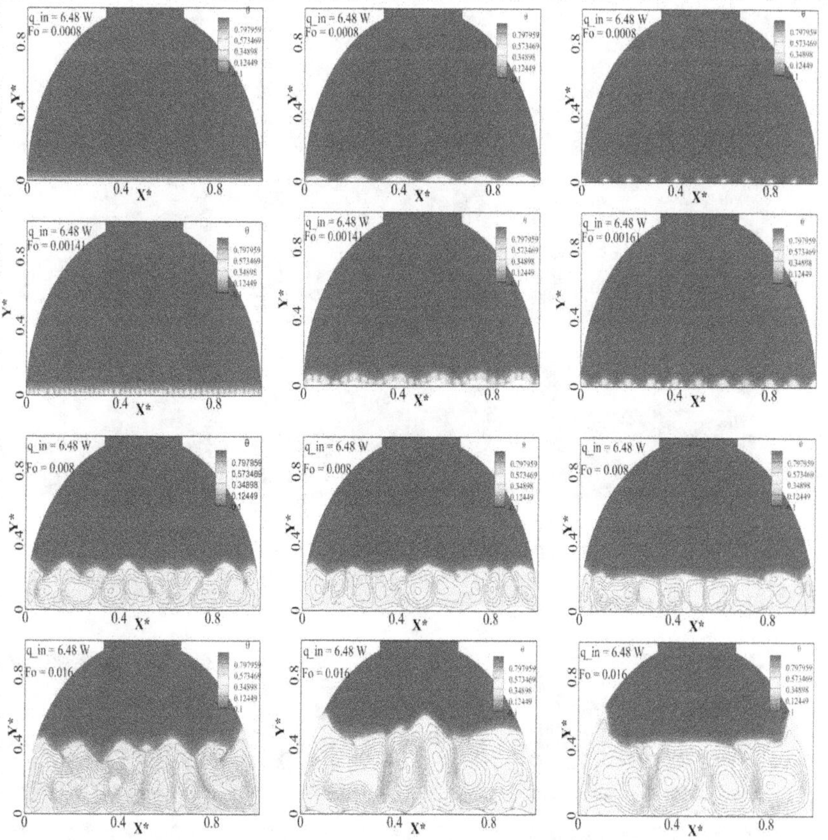

Figure 4.3: Evolution of the Temperature, superimposed with streamlines, within the semicircular enclosure for (a) straight, (b) wavy, and (c) corrugated heat-transfer surfaces under the same heat input (6.48 W) and varying F_O.

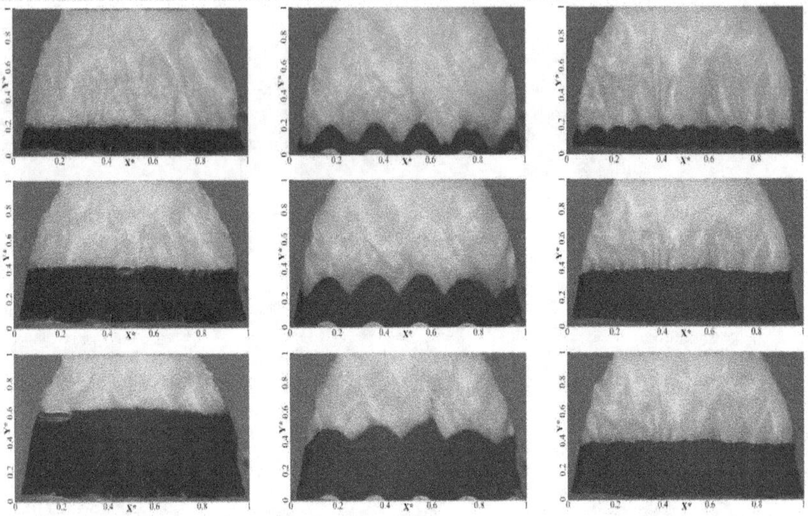

Figure 4.4: Images from experiments to evaluate melt fraction at a heat input of (5.76 W) for Fo = 0.006, 0.012, and 0.018 (arranged vertically) for (a) flat, (b) wavy, and (c) corrugated surfaces.

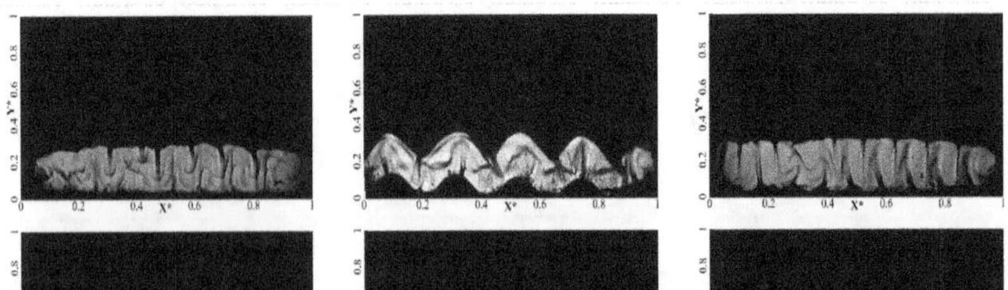

Figure 4.5: Comparison of the experimental visualization of plume evolution across three heat-transfer configurations: (a) flat surface, (b) wavy surface, and (c) corrugated surface.

4.2 Heat transfer characteristics

Convection currents formed within the molten PCM are associated with an increase in the vertical velocity component, with its direction alternating between upwash and downwash, as summarized in Figure. 4.6. The contours of the vertical velocity component reflect a similar evolution of plume structures, making them a suitable indicator for analyzing plume formation in each configuration. The number of plumes is determined through image analysis by detecting variations in pixel intensity along the horizontal direction.

It is observed from Figure 4.7 the number of plumes identified from the Y-velocity contours shows a gradual decrease with time across all configurations. Initially, multiple small plumes emerge due to strong buoyancy-driven instabilities near the heated surface. As melting progresses and the liquid layer thickens, these plumes begin to merge, weaken, or dissipate, resulting in fewer but larger convection structures at later stages. This trend is consistently observed for all inclination angles and surface roughness conditions.

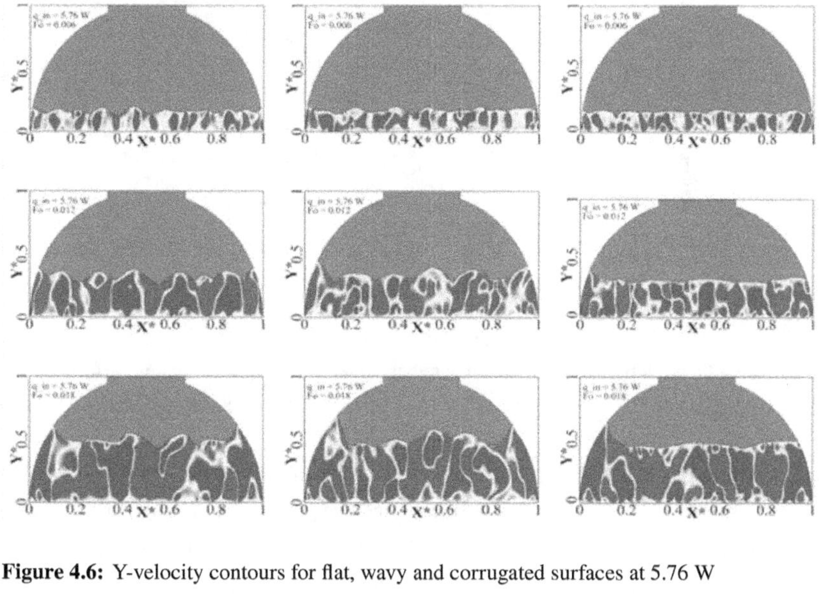

Figure 4.6: Y-velocity contours for flat, wavy and corrugated surfaces at 5.76 W

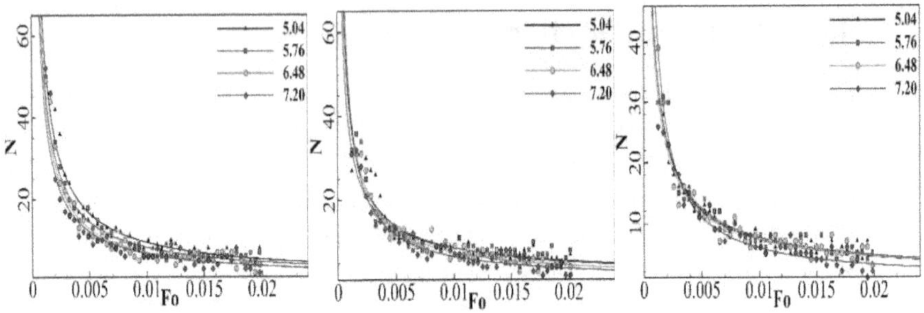

(a) (b) (c)

Figure 4.7: Evolution of plume structures under different heat inputs for the three heating surface configurations: (a) flat surface, (b) wavy surface, and (c) corrugated surface.

4.3 Scaling non-dimensional parameters, and correlations

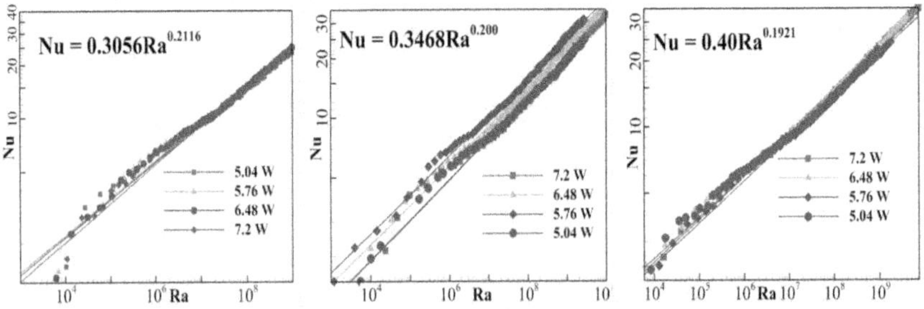

(a) (b) (c)

Figure 4.8: Evolution of the Nusselt number under various heat transfer surface configurations.(a) Flat surface, (b) wavy surface, and (c) corrugated surface.

The positive exponent m ≈ 0.20 signifies that Nu (a dimensionless heat transfer coefficient) grows sublinearly with Ra (Figure 4.8)i.e., as melt height h increases and buoyancy ramps up (higher Ra), the effective convective "efficiency" rises, but not explosively (unlike turbulent regimes where m > 0.3). This reflects the boundary layer's self-similar adjustment (Figure4.9) : stronger convection thins the thermal layer (as seen in $\delta_T \sim Ra^{-0.2}$), steepening wall temperature gradients to sustain constant heat inputs.The obtained δ_T~Ra and Nu-Ra results show good agreement with the correlations reported by Sparrow and Gregg [91, 92].

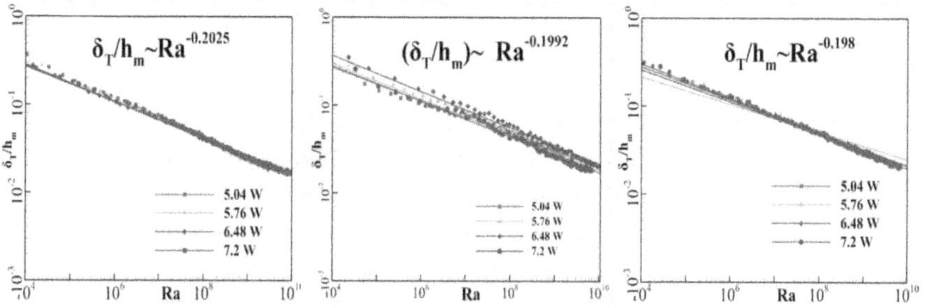

(a) (b) (c)

Figure 4.9: Evolution of the non-dimensional boundary layer thickness with respect to Rayleigh number for(a) flat, (b) wavy, and (c) corrugated surfaces.

Table 4.1: Error estimates for the correlation $Nu = C \cdot Ra^m$ (semicircular enclosure)

Heat transfer surface configuration	Heat transfer rate (W)	Correlation coefficients		
		C	m	R^2
Straight	5.04	0.3300	0.2057	0.9918
	5.76	0.3205	0.2096	0.991
	6.48	0.3033	0.2125	0.991
	7.2	0.3224	0.2185	0.9856
Wavy	5.04	0.2925	0.2032	0.988
	5.76	0.4270	0.1962	0.997
	6.48	0.3722	0.1976	0.996
	7.2	0.2949	0.2033	0.988
Corrugated	5.04	0.4400	0.1869	0.989
	5.76	0.3900	0.1928	0.9896
	6.48	0.4230	0.1930	0.996
	7.2	0.3980	0.1957	0.9954

The thermal boundary layer thickness exhibits a decreasing trend with increasing Rayleigh number for all investigated heating surfaces, indicating that enhanced buoyancy forces promote stronger convective motion and facilitate the removal of heated PCM from the vicinity of the heating surface. The flat surface shows a slightly higher dependence on Rayleigh number, as indicated by the larger magnitude of the scaling exponent ((m=-0.21)), compared with the wavy ((m=-0.20)) and arc-corrugated ((m=-0.1991)) surfaces. This suggests that the thermal boundary layer over the flat surface is more sensitive to changes in buoyancy strength.

For the corrugated surfaces, the thermal boundary layer development is influenced not only by buoyancy but also by the surface geometry. The wavy and arc-corrugated surfaces promote preferential thermal plume formation at the crest and apex regions, respectively. These localized plume structures enhance convective transport and modify the thermal boundary layer distribution along the heating surface. Consequently, further increases in Rayleigh number produce a comparatively smaller reduction in boundary layer thickness, resulting in a slightly lower magnitude of the scaling exponent.

It should also be noted that, under constant heat input conditions, the effective heat flux differs among the surfaces due to variations in the actual heated area. The corrugated surfaces possess a larger heating area compared with the wavy and flat surfaces, resulting in a lower heat flux for the same supplied power. Despite this lower heat flux, the corrugated surfaces exhibit enhanced convective activity due to the geometry-induced modification of thermal plume development and improved interaction between the heated surface and the surrounding PCM. This indicates that the influence of surface geometry plays an important role in augmenting heat transfer during PCM melting.

4.3 Summary

In summary, this study provides a comprehensive experimental and numerical assessment of Rayleigh-Bénard convection in a semi-circular enclosure under constant heat flux conditions. The influence of bottom surface geometry on convective structure, melt front evolution, and heat transfer characteristics is clearly demonstrated. Flat heating surfaces yield the highest average Nusselt number, while corrugated surfaces exhibit the lowest performance. The proposed Nu-Ra and thermal boundary layer scaling correlations capture the unique melting transport physics of curved enclosures. These findings offer valuable guidance for the design and optimization of latent heat storage systems and PCM-based electronic cooling applications.

5 Effect of unsteady heating patterns in melting dynamics and heat transfer

This chapter presents a detailed investigation of melting dynamics and heat transfer over wavy, flat and corrugated surfaces for latent heat energy storage applications. A comprehensive experimental as well as numerical analysis of different surface configurations operating under varying heat inputs is carried out. The flow characteristics within the 2-D semicircular enclosure are examined, including the number of plumes and the associated Y-velocity contours. Additionally, empirical Nu-Ra correlations are proposed to predict the influence of surface geometry on heat-transfer performance.

5.1 Introduction

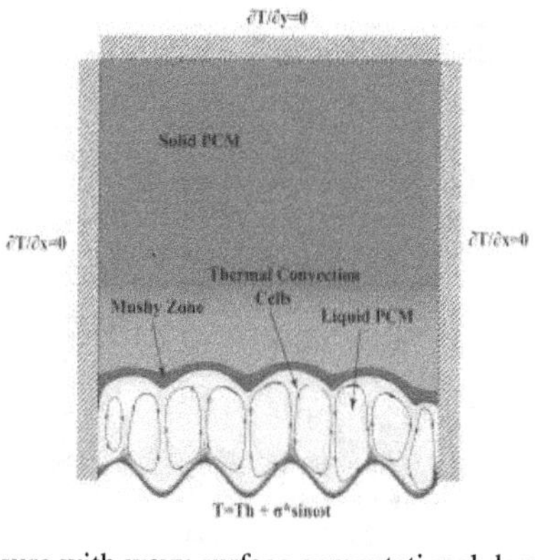

Figure 5.1: Square enclosure with wavy surface computational domain with boundary conditions

---

A detailed presentation of the results presented in this chapter has been published as Osman, M.F. and Deepu, M., 2025. Effects of Rapid Boundary Heat Flux Fluctuations on Wavy Heat-Transferring Surfaces in Latent Energy Storages. Journal of Thermal Science and Engineering Applications, 17(7), p.071003. doi :https://doi.org/10.1115/1.4068343

In Chapter 4, the melting dynamics of a PCM inside a semicircular enclosure under uniform heat input conditions for three distinct heating surfaces was investigated. However, no studies on melting dynamics for wavy surfaces under cyclic heating conditions have been published; thus, the focus of this chapter will be melting dynamics of a PCM with a wavy heating surface under cyclic heat input. The bottom wavy surface profile is defined as $Asin(\pi x/\lambda)$, where "A" represents the wave amplitude and λ denotes the wavelength. Multiple configurations with varying amplitude-to-wavelength ratios (AWR) ranging from 0.5 to 2 have been examined in this study. A periodic temperature boundary condition is applied to the wavy surface, defined as T(t)=$\overline{T}$+σsin(ωt), where $\overline{T}$ is the mean imposed temperature (350 K).The two vertical and upper horizontal walls are assumed to be adiabatic. The entire study in this chapter is divided into two sections. In the first half of the analysis, the amplitude(σ) is varied from 0 K to 40 K, while the frequency remains constant (1 rad/s). In the second portion of the analysis, the frequency (ω) is varied from 0 to 4 rad/s while the amplitude (σ) remains constant (σ=10K). The mean imposed temperature ($\overline{T}$) is chosen so that it does not enable the temperature to drop far below the solidus temperature of the PCM (316.35 K).

5.1 Time period (s) estimates for different amplitudes (ω = 1 rad/s)

		Amplitude (σ (K))				
Melt height	AWR	0	10	20	30	40
2.5	0.5	21.26	20.32	19.02	18.72	18.31
	1	21.34	20.88	20.04	19.66	19.05
	1.5	22.92	22.32	21.87	21.44	21.13
	2	24.94	23.12	22.77	22.23	22.09
5	0.5	24.63	18.2	16.05	15.76	15.38
	1	23.96	19.12	18.44	17.98	17.44
	1.5	24.84	22.34	21.68	21.13	20.85
	2	25.61	23.01	22.04	21.58	21.15
10	0.5	48.86	17.28	15.91	15.48	15.04
	1	37.14	18.18	16.98	16.66	16.31
	1.5	32.87	19.52	18.26	17.96	17.54
	2	28.01	20.33	19.24	18.84	18.34

5.2 Time period (s) estimates for different frequencies (σ = 10 K)

Frequency (ω (rad/s))				
0	10	20	30	40
21.26	20.32	19.02	18.72	18.31
21.34	20.88	20.04	19.66	19.05
22.92	22.32	21.87	21.44	21.13
24.94	23.12	22.77	22.23	22.09
24.63	18.2	16.05	15.76	15.38
23.96	19.12	18.44	17.98	17.44
24.84	22.34	21.68	21.13	20.85
25.61	23.01	22.04	21.58	21.15
48.86	17.28	15.91	15.48	15.04
37.14	18.18	16.98	16.66	16.31
32.87	19.52	18.26	17.96	17.54
28.01	20.33	19.24	18.84	18.34

5.2 Augmentation of Rayleigh-Bénard convection with the fluctuating heat input

Although melting and solidification in a phase-changing medium may appear straightforward to a casual observer, scientific investigations reveal a highly intricate interaction of coupled transport phenomena involving heat diffusion, phase change, and Rayleigh-Bénard convection. A deeper understanding of melt evolution is therefore crucial, especially when examining the influence of time-varying heat input on Rayleigh–Bénard convection. During the initial stage, the melt layer remains thin and nearly parallel to the heating surface, preventing the onset of instability; heat diffusion dominates at this point due to steep thermal gradients. As melting progresses, convection gradually becomes the primary mode of heat transfer. Prior studies have shown that numerous convection cells emerge at the onset of instability and subsequently merge into larger cells as the melt front recedes from the heated surface.

With time-varying heat input, notable changes occur in the morphology of convection cells forming between the phase boundary and the heating surface. These transformations arise from heat-induced temperature gradients, their interaction with fluid motion, and the temperature dependence of thermophysical properties. A wavy surface accommodates these convection cells over a larger portion of the melt height. To explore this behavior, numerical simulations are conducted to evaluate the time period of convection currents under various unsteady thermal inputs. Figure 5.2 illustrates the simulated trajectories of a particle within the convection cell for a wavy surface with AWR = 1 subjected to a fluctuating heat input of amplitude σ = 10 K and frequency ω = 1 rad/s at different melt heights. As the melt front advances, convection cells expand and merge, causing the particle's circulation time to increase. The particle velocity is reduced near the wavy surface and the phase boundary due to strong velocity gradients.

To further characterize the convection cells over wavy surfaces, their time periods are quantified for different amplitude-to-wavelength ratios (0.5 $\leq$ AWR $\leq$ 2.0) under a range of fluctuating thermal inputs with frequencies (0 rad/s $\leq \omega \leq$ 4 rad/s) and amplitudes (0 K $\leq \sigma$ $\leq$ 40 K). The results are summarized in Tables 5.1 and 5.2. For all wavy surfaces studied, the convection-cell time period increases with melt height, consistent with the growth of convection cells. Low-AWR surfaces generate vigorous convection during the early melting stages, leading to decreasing time periods as melt height increases. In contrast, high-AWR surfaces produce strong plumes that anchor near the crest of the waves and intensify as melting progresses; hence, their time periods decrease more significantly with melt height compared to lower-AWR surfaces. The time period further decreases with increasing amplitude of surface temperature oscillation, indicating enhanced convection. Melt fraction evolution also supports this observation. However, variations in temperature frequency have minimal influence on the time period due to the inherently low thermal diffusivity of the PCM.

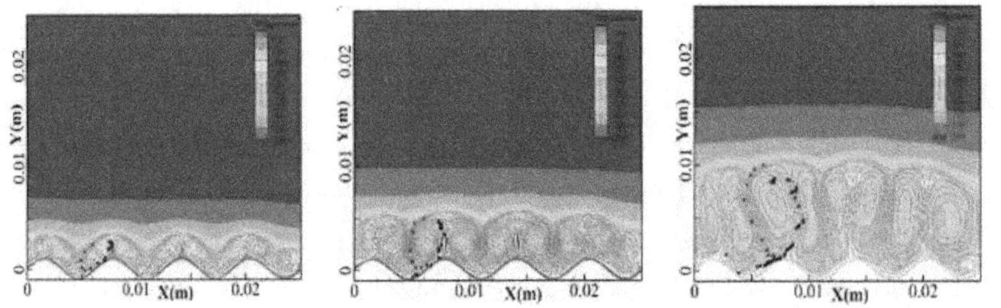

(a) (b) (c)

Figure 5.2: Visualisation of particle trajectories within the convection cell, shown alongside streamline and temperature contours, for a wavy surface with AWR = 1 subjected to a fluctuating heat input of amplitude σ = 10 K and frequency ω = 1 rad/s at different melt heights: (a) 2.5 mm, (b) 5 mm, and (c) 10 mm.

5.3 Effect of amplitude to wavelength ratio of the heat transferring boundary

The temporal evolution of the convective flow structure within the PCM confined by a wavy heating surface subjected to a time-varying heat input (frequency ω = 1 rad/s and amplitude σ = 20 K) is illustrated in Fig. 5.3. The aim of this analysis is to examine the behavior of the phase interface and its modifications under different heating-surface geometries when exposed to an oscillating thermal boundary condition. Dimensionless temperature contours superimposed on streamlines depict the flow patterns and regions of thermal accumulation.

As in the fixed-temperature boundary condition case, the time-varying heat input initially produces a melt interface parallel to the heating surface. Once instability sets in, multiple convection cells appear and gradually merge as the melt height increases, driven by strong thermal gradients that induce buoyancy-driven convection within the liquid PCM. The figure also shows that the amplitude of interface undulations decreases with a reduction in the amplitude-to-wavelength ratio (AWR).

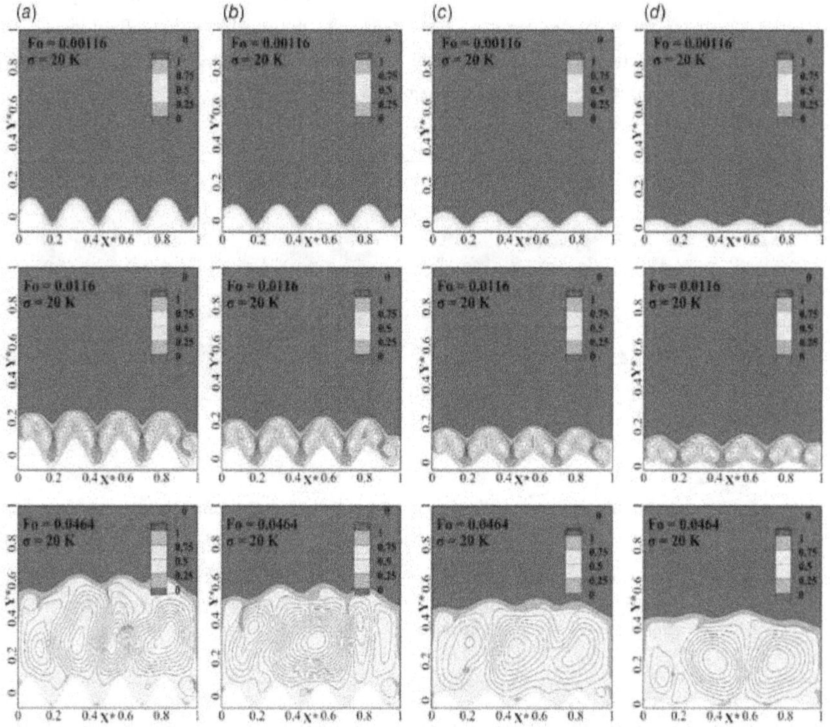

Figure 5.3: Evolution of heat convection within the molten PCM for different amplitude-to-wavelength ratios, shown through dimensionless temperature contours superimposed with streamlines for a case subjected to a time-varying heat input of frequency ω = 1 rad/s and amplitude σ = 20 K.

At moderate melt heights, high-temperature plumes rise from the crests of the wavy boundary, while the downward-moving cooler fluid impinges on the troughs, forming convection cells approximately equal in size to the surface wavelength. As melting progresses, surfaces with higher AWR develop greater curvature in the melt interface and exhibit enhanced melt fractions. The sharp crests of high-AWR surfaces anchor strong upward-moving thermal plumes, which transfer heat more effectively to the phase boundary during the coarsening and coalescing stages of convection.

Convective activity intensifies with increasing melt height, enabling the hot plume to oscillate near the interface and improve heat transfer to the mushy region. Regions of intense convection are particularly evident near the receding melt interface, whereas heat diffusion becomes dominant in the vicinity of the heating surface.

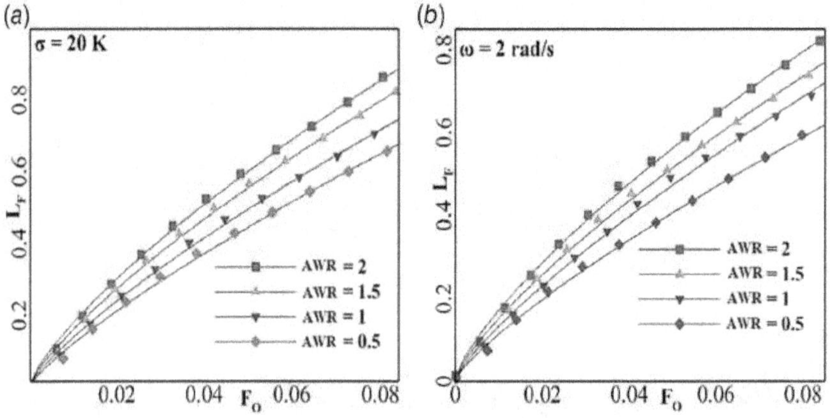

Figure 5.4: Melt fraction evolution over time for a case with time-varying heat input at amplitude σ = 20 K, shown for (a) ω = 1 rad/s and (b) ω = 2 rad/s, across various amplitude-to-wavelength ratios (0.5 $\leq$ AWR $\leq$ 2).

Figure 5.4 shows the temporal variation of melt fraction with non-dimensional Fourier number (F_O). Higher surface waviness consistently results in a larger melt fraction under oscillating thermal input because steeper temperature gradients generate stronger instabilities. Although the influence of waviness remains similar across different frequencies, a substantial change in melt fraction with increasing frequency is not observed, owing to the PCM's low thermal diffusivity. Moreover, a nonlinear melt-fraction evolution with time is observed for fluctuating boundary conditions, contrasting with the linear melt-fraction development reported by Madruga and Curbelo [93] for steady-temperature heating.

5.4 Effect of Amplitude and Frequency of the Fluctuating Heat Input at the Heat-Transferring Boundary in Melting

A study is conducted to examine the influence of amplitude and frequency of fluctuating heat input at the wavy heat-transferring boundary. First, the melting behavior and phase morphology of the PCM inside the enclosure are analyzed for a fixed frequency of ω = 1 rad/s and varying amplitudes (0 K < σ < 40 K) across different amplitude-to-wavelength ratios (0.5 < AWR < 2). Figure 5.5 shows the evolution of the melt fraction at F_O = 0.0348 for various AWR values. Earlier studies have shown that higher AWR leads to greater waviness in the phase interface. The present results further reveal that the waviness of the interface increases with higher amplitude of temperature fluctuations. For high-AWR surfaces, the hot plume anchored at the crest exhibits larger oscillations, generating stronger gradients at the interface. This interaction creates a downward-extending mushy trail within the melt pool; the semisolid mush periodically detaches and melts as it is carried by the convection currents. The strength of this transport mechanism intensifies with increasing amplitude. The enhancement in interface curvature becomes even more pronounced for lower-AWR surfaces.

Figure 5.6 presents the temporal variation of melt fraction. As established previously, surface waviness enhances melting under fluctuating boundary conditions. Additionally, increasing the amplitude of temperature oscillation significantly improves melt fraction at all AWR levels. This enhancement becomes more prominent as melting continues, especially for high-AWR surfaces subjected to larger amplitudes. These observations confirm the strong influence of temperature-amplitude variations in promoting melt progression and heat transfer in enclosures with wavy heating boundaries.

Across all cases, the L_F-F_O relation follows a power law with an exponent ranging from 0.72 to 0.83 and R^2 = 0.99. Table 5.1 summarisess R^2 values for the two conditions examined: constant amplitude and constant frequency. In both conditions, the lowest exponent corresponds to AWR = 2, while the highest is associated with AWR = 0.5.

In the next stage, melting behavior and PCM morphology are evaluated for a fixed amplitude of σ = 10 K and varying frequencies (0 rad/s < ω < 4 rad/s) across different AWR values. Figure 5.7 illustrates the melt fraction distribution at F_O = 0.0348 for different waviness levels. Heat input at ω = 1 rad/s induces more pronounced oscillations at the melt interface, particularly for surfaces with lower AWR. Further analysis with fixed amplitude (σ= 10 K) confirms that ω = 1 rad/s consistently produces the highest melt fraction throughout the melting process for all surface configurations. This highlights the role of the PCM's diffusion time in determining how effectively boundary-temperature fluctuations influence melting, as supported by Figures 5.6, 5.7 and Table 5.1.

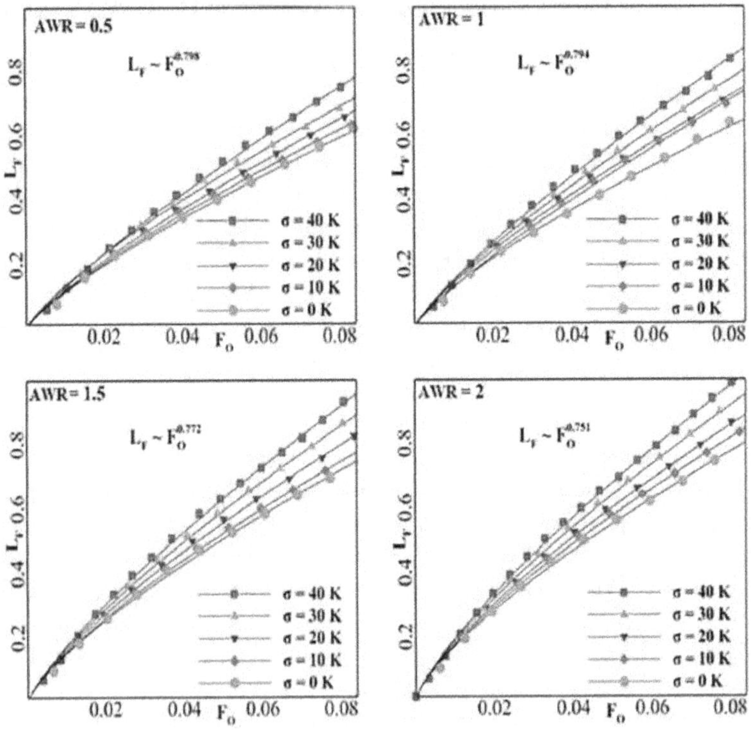

Figure 5.5: Melt fraction evolution under a time-varying heat input with fixed frequency, ω = 1 rad/s and varying amplitudes (0 K < σ < 40 K) for different amplitude-to-wavelength ratios (0.5 < AWR < 2) of the heating surface.

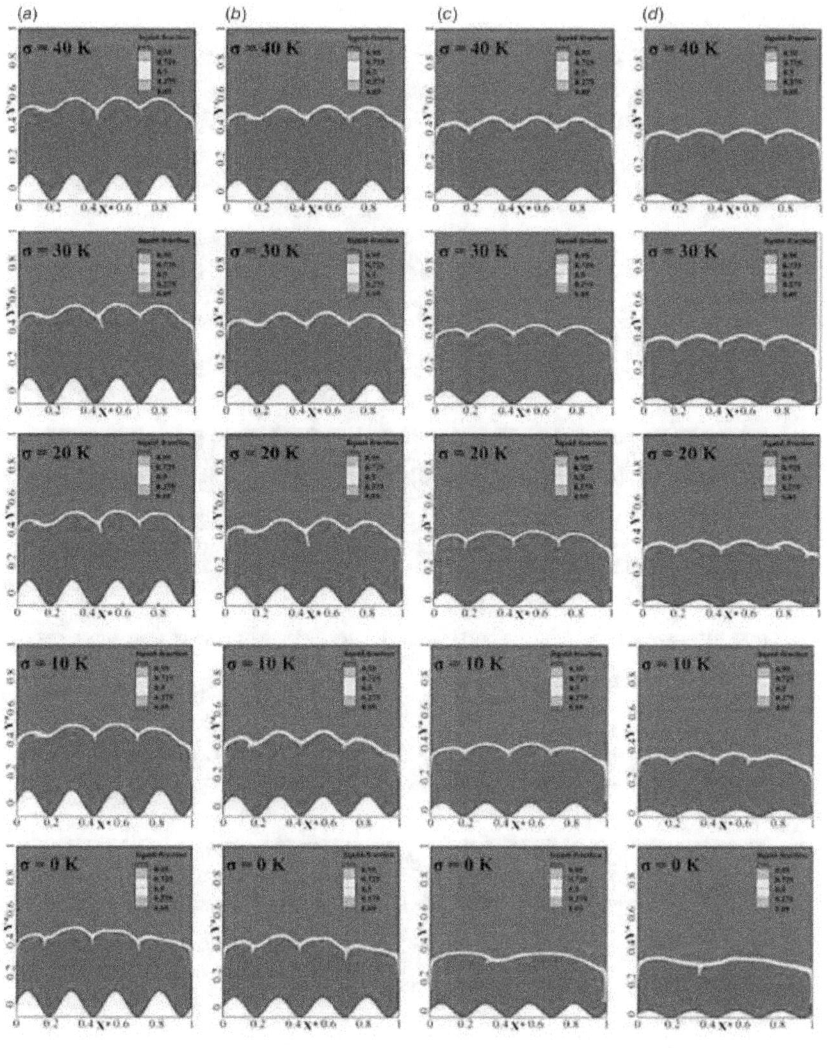

Figure 5.6: Evolution of the melt fraction within the enclosure at F_O = 0.0348 for cases subjected to time-varying heat input with a fixed frequency of ω = 1 rad/s and varying amplitudes (0K < σ < 40K).

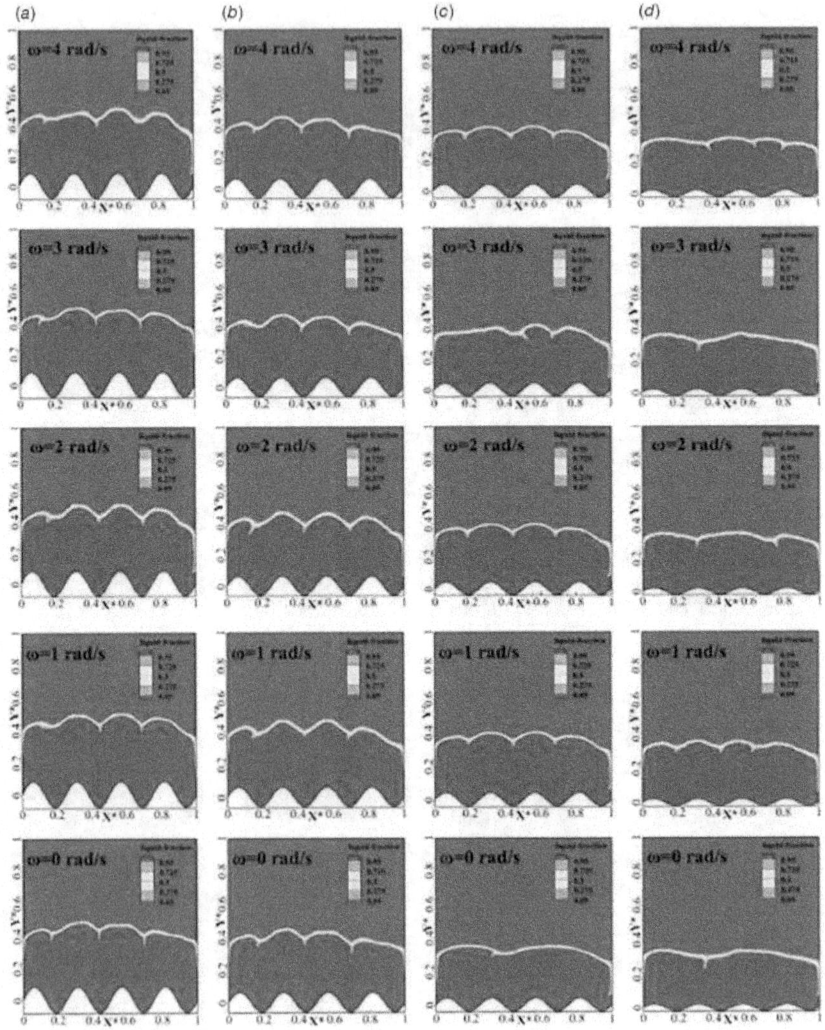

Figure 5.7: Melt fraction evolution within the enclosure at F_O = 0.0348, under oscillatory heating with fixed amplitude(σ =10 K) and and variable frequencies (0 rad/s≤ω ≤ 4 rad/s)

Table 5.1: Power law relation factors for L_F Vs F_O curves

		Constant Amplitude (σ = 10 K)			Constant Frequency (ω = 1 rad/s)			
Configuration	ω (rad/s)	p	q	R^2	Amplitude(σ(K))	p	q	R^2
AWR 0.5	0	4.00282	0.7677	0.99831	0	4.00282	0.7677	0.99831
	1	4.58138	0.7958	0.99784	10	4.59974	0.7964	0.99769
	2	4.31458	0.7912	0.99869	20	4.97288	0.8063	0.99812
	3	4.41055	0.7881	0.9985	30	4.94808	0.7817	0.99776
	4	4.19965	0.7804	0.99858	40	6.21521	0.8386	0.99841
AWR 1	0	4.20385	0.7576	0.99836	0	4.20385	0.7576	0.99836
	1	5.65195	0.7951	0.99792	10	5.6576	0.788	0.9979
	2	5.01283	0.7871	0.99759	20	5.2488	0.7987	0.99826
	3	4.88906	0.7815	0.99736	30	5.78923	0.8001	0.99868
	4	4.52221	0.7681	0.99914	40	6.26513	0.8235	0.99866
AWR 1.5	0	4.76358	0.75	0.99873	0	4.76358	0.75	0.99873
	1	5.03795	0.758	0.99886	10	5.05309	0.7585	0.99876
	2	4.96295	0.7557	0.99845	20	5.55668	0.7697	0.99866
	3	4.78746	0.7511	0.99899	30	6.17803	0.7833	0.99845
	4	4.78267	0.7508	0.99887	40	6.87574	0.7985	0.99888
AWR 2	0	4.84041	0.7278	0.99875	0	4.84041	0.7278	0.99875
	1	5.26	0.74	0.9988	10	5.27511	0.7403	0.9987
	2	5.12	0.7366	0.99863	20	5.6576	0.7491	0.99887
	3	4.933	0.7302	0.99884	30	6.25262	0.7617	0.99896
	4	4.81	0.7258	0.9987	40	6.93097	0.7767	0.99894

5.5 Scaling Laws and Heat-Transfer Correlations for Melt-Height Evolution Under Fluctuating Heat Input at a Wavy Thermal Boundary

The progression of melt height is a key parameter in latent heat storage analysis, as it directly reflects the energy storage capacity and thermal response of the system. Consequently, it plays a central role in evaluating the performance and guiding the design of latent heat storage units. While Madruga and Curbelo [93] reported a linear liquid-fraction evolution dependent on domain size, the present study demonstrates a nonlinear melt fraction behavior driven by the imposed periodic thermal boundary condition.

Figure 5.8 illustrates the dimensionless melt height (h*) evolution with dimensionless time

(F_O) for various amplitude-to-wavelength ratios ($0.5 \leq$ AWR ≤ 2) under a fluctuating heat input with fixed frequency ω = 1 rad/s and varying amplitudes (0 K $\leq \sigma \leq$ 40 K). The melt height follows a power-law trend with an exponent greater than 0.5, consistent with diffusion-dominated Stefan problems. Results show that higher AWR surfaces tend to shift more toward diffusion-controlled behavior. For any given AWR, the largest amplitude of temperature oscillation produces the highest exponent, indicating enhanced melting.

A similar analysis performed for fixed amplitude σ = 10 K and varying frequencies (0 rad/s $\leq \omega \leq$ 4 rad/s) shows that the melt height evolution is only weakly dependent on frequency. This observation aligns with earlier results on convection structure evolution (Fig. 5.5), where fewer streamlines are produced near the heating surface for high-AWR cases.

Figures 5.10 and 5.11 present the variation of Nusselt number with Rayleigh number for different test conditions. The obtained Nu-Ra exponents are slightly less than the classical 2/7 exponent value reported for PCM melting over flat surfaces [6]. Its observed from the Fig 5.10. For ω = 0 rad/s, the absence of temporal forcing first establishes steady, balanced convection, after which plumes remain stationary and lead to only moderate thermal stratification without significant enhancement. At ω = 1 rad/s (τ=6.28 s), the imposed oscillation first introduces rhythmic stirring of the melt pool, with hot phases generating vigorous, deep-penetrating plumes and cold phases suppressing excessive stratification. This interaction subsequently enhances recirculation by 20–40% through Dean-like instabilities induced by the wavy geometry, increasing solid–liquid interface exposure and accelerating latent heat absorption. For ω = 4 rad/s (τ=1.57 s), the oscillations first occur too rapidly relative to plume growth time, causing buoyancy-driven motions to be damped by viscosity before fully developing. As a result, flow remains confined to shallow near-wall oscillations, leading to slower bulk melting and the lowest melt fraction, since energy is repeatedly spent on boundary-layer cycling without advancing the melting front. Fig. 5.11 shows Nusselt number evolution for varying amplitudes. For AWR = 2, the highly undulated wavy surface with pronounced peaks and troughs first induces strong recirculation zones and disrupts the thermal boundary layer, leading to enhanced mixing and higher local convective velocities near the solid–melt interface. As a result, the effective heated area increases due to the longer arc length, which further thins the thermal boundary layer and enhances convective heat transfer within the melt pool. When the heating amplitude ais non-zero, the imposed temperature oscillation first introduces pulsatile heating, in which the hot phases (T>350 K) generate stronger buoyancy-driven plumes while the cooler phases permit partial regeneration of the thermal boundary layer. With increasing AWR, the wavy surface then amplifies these oscillatory plumes, as wave-induced eddies guide hot fluid deeper into the melt domain. Consequently, the time-averaged heat flux increases, accelerating latent heat absorption and leading to a higher melt fraction. The highest exponent value "q" for AWR = 2 indicates that convection strengthens markedly with increasing buoyancy force, whereas for AWR = 0.5, the weaker surface undulations limit flow development, resulting in the least enhancement of convection with buoyancy.

Table 5.2: Power law relation factors for dimensionless height (h^*) vs dimensionless time (F_O)

		Constant Amplitude (σ = 10 K)			Constant Frequency (ω = 1 rad/s)			
Configuration	ω (rad/s)	p	q	R^2	Amplitude σ (K)	p	q	R^2
AWR 0.5	0	3.89	0.6885	0.999833	0	3.89	0.6885	0.999833
	1	4.16	0.7171	0.999902	10	4.16	0.7171	0.999902
	2	4.137	0.7118	0.999433	20	4.18	0.7254	0.999765
	3	3.975	0.7066	0.999894	30	4.26	0.7614	0.999661
	4	3.905	0.6995	0.999808	40	4.2	0.7655	0.99351
AWR 1	0	3.713	0.6357	0.998546	0	3.713	0.6357	0.998546
	1	3.777	0.6821	0.998879	10	3.777	0.6821	0.998879
	2	3.53	0.6663	0.99938	20	4.087	0.6957	0.999173
	3	3.12	0.6452	0.99905	30	4.545	0.7369	0.999967
	4	3.09	0.6406	0.999607	40	4.945	0.7405	0.998367
AWR 1.5	0	3.59	0.6129	0.999295	0	3.59	0.6129	0.999295
	1	4.087	0.6394	0.999173	10	4.087	0.6394	0.999173
	2	4.0675	0.6314	0.984837	20	3.777	0.6298	0.998879
	3	4.07368	0.6219	0.986692	30	4.67	0.6651	0.999001
	4	4.063	0.6135	0.97875	40	5.06	0.6878	0.998471
AWR 2	0	3.89	0.589	0.998284	0	3.89	0.589	0.998284
	1	4.06	0.6121	0.998393	10	4.06	0.6121	0.998393
	2	4.0730	0.6098	0.991865	20	4.216	0.624	0.999218
	3	4.07607	0.6011	0.991759	30	4.73	0.6446	0.997148
	4	4.0746	0.5956	0.990402	40	5.098	0.6527	0.998374

Table 5.3: Power law relation factors and R^2 values for $Nu = p \cdot Ra^q$ relation for melting over wavy surfaces with time-varying heat input

		Constant Amplitude (σ = 10 K)			Constant Frequency (ω = 1 rad/s)			
Configuration	ω (rad/s)	p	q	R^2	Amplitude σ (K)	p	q	R^2
AWR 0.5	0	0.4036	0.1924	0.998	0	0.1689	0.2396	0.9955
	1	0.2650	0.2062	0.9993	10	0.2153	0.2369	0.9986
	2	0.5792	0.2053	0.9976	20	0.3059	0.2386	0.9985
	3	0.5873	0.2002	0.9959	30	0.2967	0.2575	0.9998
	4	0.6954	0.1968	0.9993	40	0.3857	0.2837	0.9973
AWR 1	0	0.3738	0.19293	0.9984	0	0.0776	0.2605	0.9898
	1	0.2777	0.2108	0.9971	10	0.1718	0.2454	0.99989
	2	0.48406	0.2085	0.9996	20	0.3277	0.2525	0.99999
	3	0.5508	0.2099	0.998	30	0.3000	0.267	0.99997
	4	0.6247	0.2088	0.9993	40	0.3561	0.2902	0.99962
AWR 1.5	0	0.4339	0.1853	0.99993	0	0.1131	0.2441	0.99913
	1	0.5075	0.19687	0.99878	10	0.2104	0.2366	0.9998
	2	0.6678	0.19466	0.99945	20	0.2619	0.2429	0.99918
	3	0.6955	0.1911	0.99823	30	0.2099	0.2513	0.99967
	4	0.7464	0.1888	0.99996	40	0.3943	0.2763	0.99975
AWR 2	0	0.3740	0.1916	0.99856	0	0.1704	0.2451	0.99856
	1	0.4833	0.1944	0.99967	10	0.3204	0.2203	0.9996
	2	0.6544	0.1943	0.99976	20	0.3500	0.2348	0.99978
	3	0.7567	0.1888	0.99988	30	0.3287	0.2561	0.99173
	4	0.7597	0.1875	0.99998	40	0.5910	0.2608	0.99853

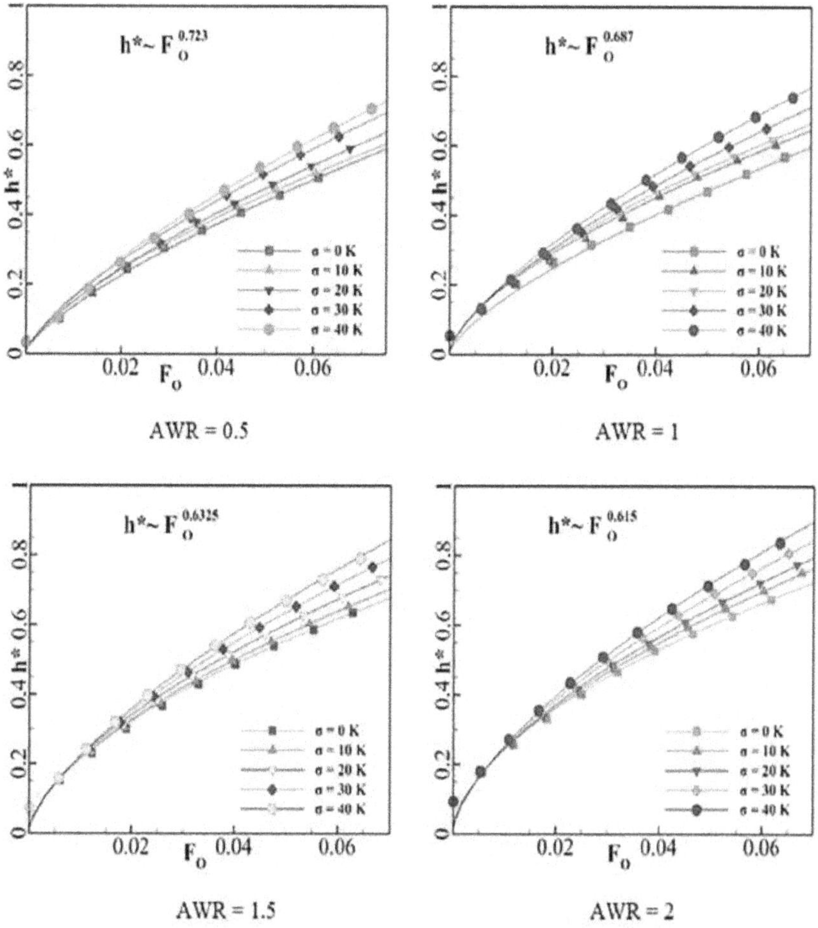

Figure 5.8: The dimensionless melt height (h*) evolution with dimensionless time (F_O) for various amplitude-to-wavelength ratios ($0.5 \leq$ AWR ≤ 2) under a time-varying heat input of fixed frequency ω = 1 rad/s and different amplitudes (0 K < σ < 40 K) is presented.

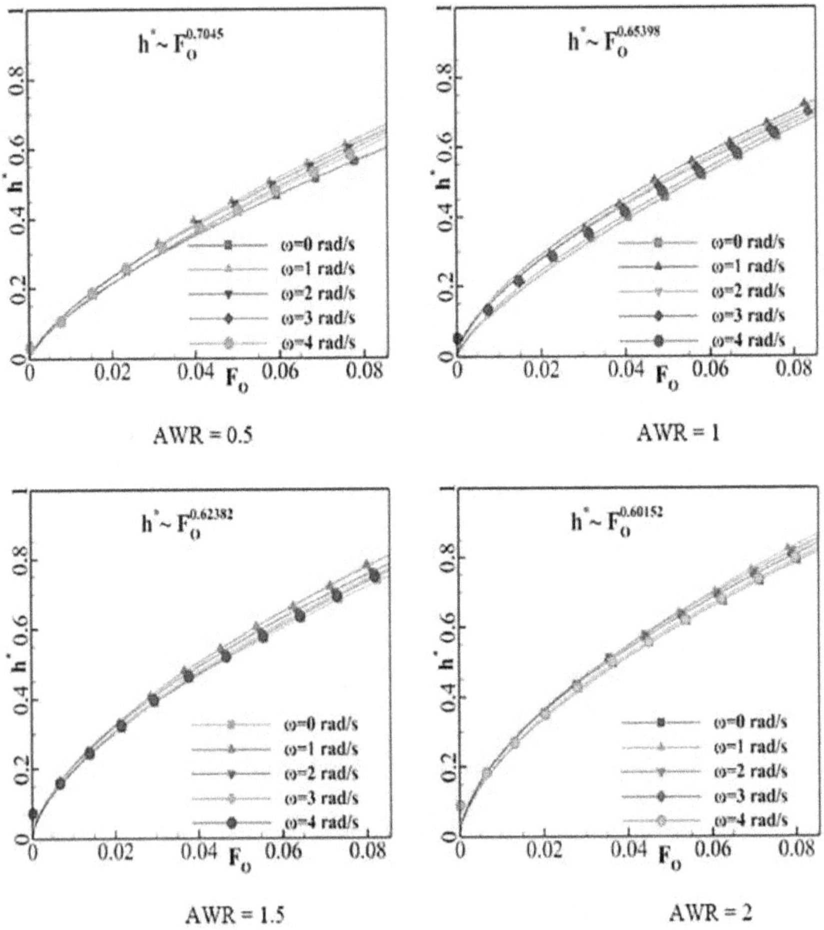

Figure 5.9: The dimensionless melt height (h*) evolution with time (F_O) for various amplitude-to-wavelength ratios ($0.5 \leq AWR \leq 2$) under a time-varying heat input of fixed amplitude σ = 10 K and different frequencies (0 rad/s $\leq \omega \leq$ 4 rad/s) is presented.

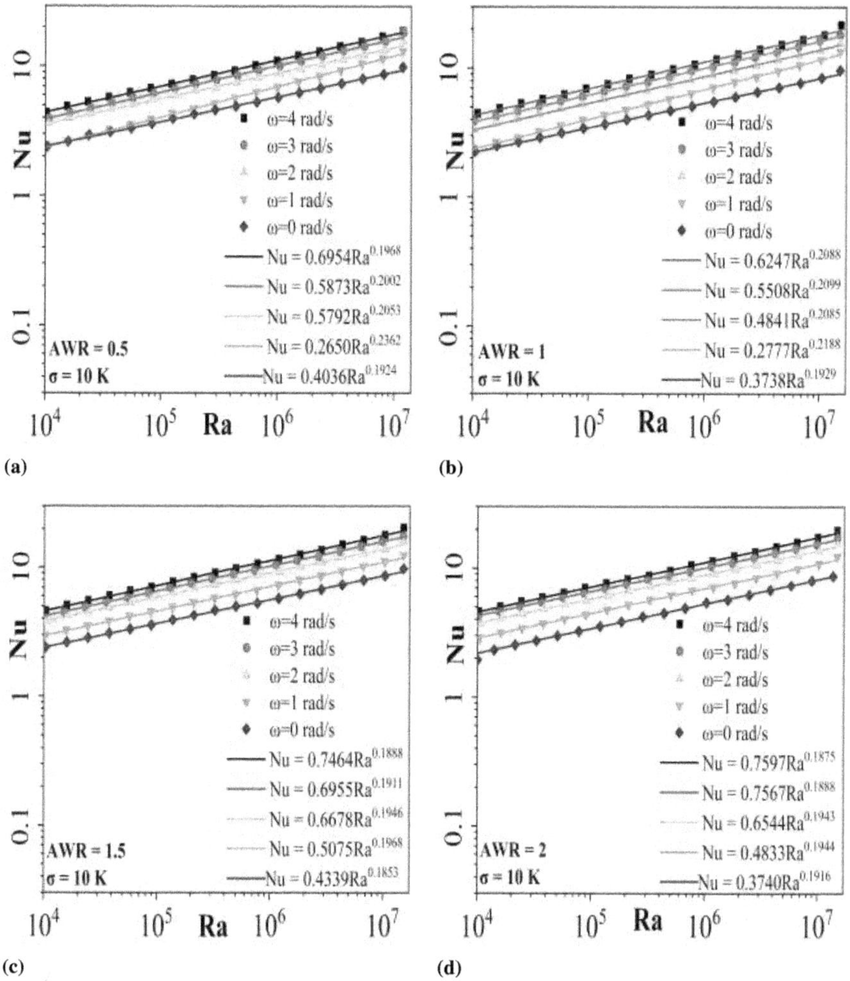

Figure 5.10: Evolution of the Nusselt number during the melting process for different amplitude-to-wavelength ratios ($0.5 \leq \text{AWR} \leq 2$) of the heating surface under time-varying heat input with a fixed amplitude (σ = 10 K) and different angular frequencies ($0 \leq \omega \leq 4$ rad/s).

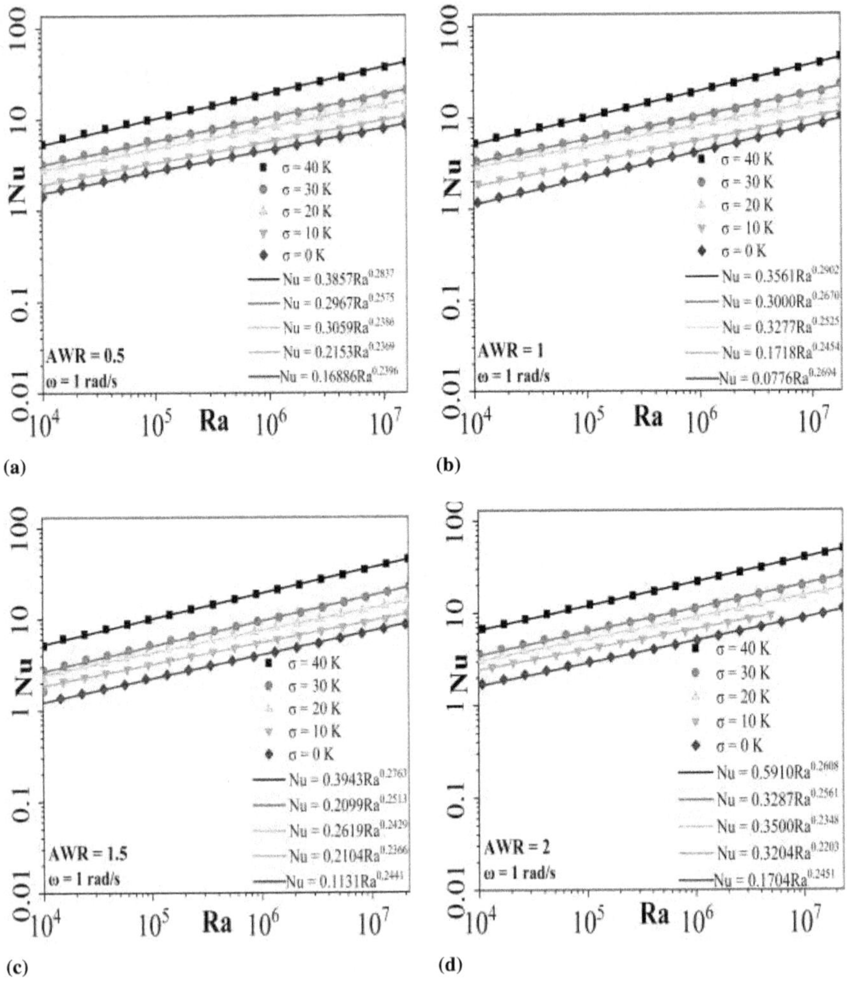

Figure 5.11: Evolution of the Nusselt number during the melting process for different amplitude-to-wavelength ratios ($0.5 \leq \text{AWR} \leq 2$) of the heating surface under time-varying heat input with a fixed angular frequencies ($\omega = 1$ rad/s) and amplitude ($0 \leq \sigma \leq 40$ K).

Higher frequency (ω) causes more rapid cycling. This prevents the boundary layer from fully stabilizing or thickening, leading to repeated disruption, reformation, and enhanced mixing. The result is a higher time-averaged heat transfer coefficient. the rapid cycling confirms to better "pump" the convection cells: hot fluid is periodically ejected upward more efficiently, and fresh cooler liquid is drawn in more frequently. This synchronization sustains stronger velocity fields over time compared to the steady case (ω= 0), where a single quasi-steady circulation pattern develops and may become less effective as the melt pool grows.

5.6 Summary

In conclusion, this study presents a detailed analysis of Rayleigh-Bénard convection during the melting of PCM over wavy heating surfaces under time-varying boundary temperatures. The results demonstrate that surface waviness and the amplitude of temperature fluctuations strongly influence convection cell morphology, melt fraction evolution, and heat transfer enhancement. Larger amplitude–wavelength ratios result in higher convective activity and faster melting rates. Nonlinear melt fraction growth and established power-law correlations ($L_F \sim F_O$, h* $\sim F_O$, Nu $\sim$ Ra) provide predictive insight for melting dynamics.

6 Effect of inclination of the PCM confinement in melting dynamics and heat transfer

This chapter presents a comprehensive experimental and numerical investigation of melting dynamics and heat transfer in a two-dimensional rectangular enclosure subjected to different inclination angles and heating surface configurations, including a flat surface, a surface with square patterned elements, and a surface with triangular patterned elements, for latent heat energy storage applications. The analysis covers various surface geometries operating under different heat inputs. Flow behavior, melt fraction evolution, and unsteady thermal fields within the enclosure are examined using temperature distributions and Y-velocity contours. In addition, empirical Nusselt-Rayleigh (Nu-Ra) correlations and time-averaged Nusselt numbers are developed for each inclination angle and heating surface configuration to quantify the impact of surface geometry on heat-transfer performance.

6.1 Introduction

The effect of uniform heat input on the melting dynamics of a PCM over wavy and corrugated surfaces is analyzed in Chapter 4, and the influence of cyclic heating with a wavy heating surface is examined in Chapter 5. However, the impact of PCM enclosure orientation in the presence of heating surface undulations on melting dynamics has not been addressed in the existing literature and is therefore investigated in this chapter.

The impact of enclosure orientation is evaluated by inclining the system from 0° and 20° to 60° in increments, with heat flux values varying between 7 kW/m^2 and 10 kW/m^2. Figure 6.1 depicts typical melting contours of a PCM under a homogeneous heat flux from the bottom and inclined at an angle (γ). Because of the density differential in the molten PCM, we see that more of the region is melted on the top than on the bottom. The top section is exposed to the environment, whereas both vertical sides are adiabatic.

A detailed presentation of the results is available in the manuscript
Osman, M.F., Vijayan, A. and Deepu, M., 2026. Effect of inclination of patterned heating surface in melting dynamics and heat transfer augmentation. Journal of Energy Storage, 160, p.121924. https://doi.org/10.1016/j.est.2026.121924

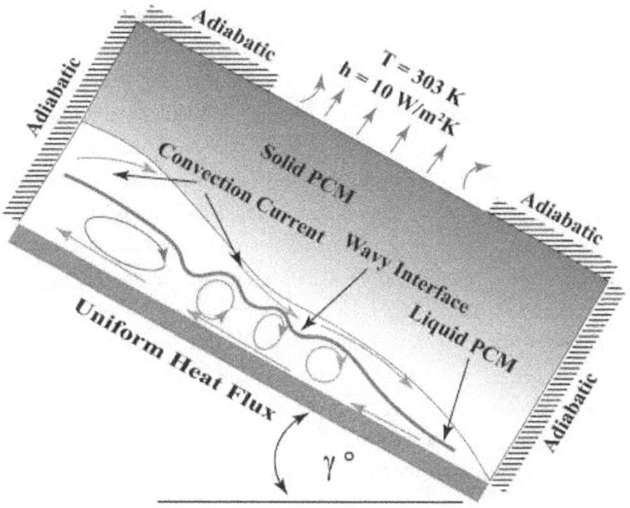

Figure 6.1: Rayleigh–Bénard convection in phase change material within a rectangular enclosure featuring flat surface and having inclination

6.2 Meltfront evolution and thermal convection physics in inclined latent heat storages

This section provides a detailed discussion of the evolution of convective flow structures within PCM enclosures featuring patterned heat-transfer surfaces at different inclination angles. Heat is supplied directly to the PCM through the heated boundary. Experimental visualizations of plume development for three heating-surface configurations at 0° inclination, captured at different time instants, are compared in Fig. 6.2. During the initial stage of melting, heat transfer is dominated by diffusion, closely resembling the one-dimensional Stefan melting behavior. As melting progresses and the liquid layer thickens, thermal instabilities emerge, leading to the formation of convection cells. Initially, numerous small convection cells develop and subsequently coalesce as the melt front moves away from the heating surface. The evolution of these convection patterns is illustrated at three representative non-dimensional times corresponding to Fourier numbers F_O = 0.0058, 0.0116, and 0.0174. The shape of the solid–liquid interface closely follows the evolving convection structures. The presence of patterned heating surfaces is observed to enhance melting, primarily due to the anchoring of hot plumes on the extended surface features. A detailed quantification of melt volume is presented in subsequent sections. Figure 6.3 shows the superimposed numerical streamlines and temperature contours for horizontal PCM enclosures heated using flat, triangular, and square roughened surfaces, exhibiting trends consistent with experimental observations.

In a horizontal enclosure, a typical convection cell consists of a hot plume rising from the heated surface and a cooler stream descending after transferring heat to the solid PCM above. However, inclining the enclosure significantly alters the convective melting behavior, as illustrated in Fig.6.4. Inclination intensifies plume oscillations due to the combined effects of buoyant rise of low-density hot melt and the downward motion of cooler melt along the inclined heating surface, thereby disrupting the vertical alignment of convection currents. This interaction results in an oscillatory interface between relatively hot and cold molten regions, along which the phase boundary progressively recedes. For the flat heating surface, the absence of surface irregularities allows the thermal boundary layer to develop smoothly along the heated wall. Initially, the heated liquid layer near the surface becomes unstable due to buoyancy forces, leading to the formation of thermal plumes. However, because there are no geometrical disturbances to alter the flow, the plumes originate at relatively predictable locations and rise in an organized manner. This results in a relatively orderly and stable plume motion. The thermal boundary layer remains comparatively uniform, and the interaction between neighboring plumes is weaker.

For the surface with square patterned elements, the sharp-edged square protrusions act as flow disturbances and additional nucleation sites for plume generation. The presence of 30 square elements with a 1.92 mm gap between adjacent elements creates repeated acceleration and separation regions near the heating surface. As the heated fluid encounters the vertical faces and corners of the square elements, the thermal boundary layer is repeatedly disrupted, causing local variations in temperature gradients and buoyancy forces. The sharp corners generate stronger flow separation and recirculation zones, which promote frequent plume formation, merging, and detachment. As a result, the plume motion becomes highly irregular and vigorous compared to the flat surface. For the triangular-patterned surface, the inclined faces of the triangular elements also disturb the thermal boundary layer; however, the disturbance is smoother compared to the abrupt flow separation produced by square elements. The sloping surfaces guide the upward movement of heated fluid and promote plume generation, but the absence of sharp corners reduces intense flow separation and recirculation. Therefore, the plume behavior is more active than the flat surface but less chaotic than the square-patterned surface, resulting in an intermediate plume motion.

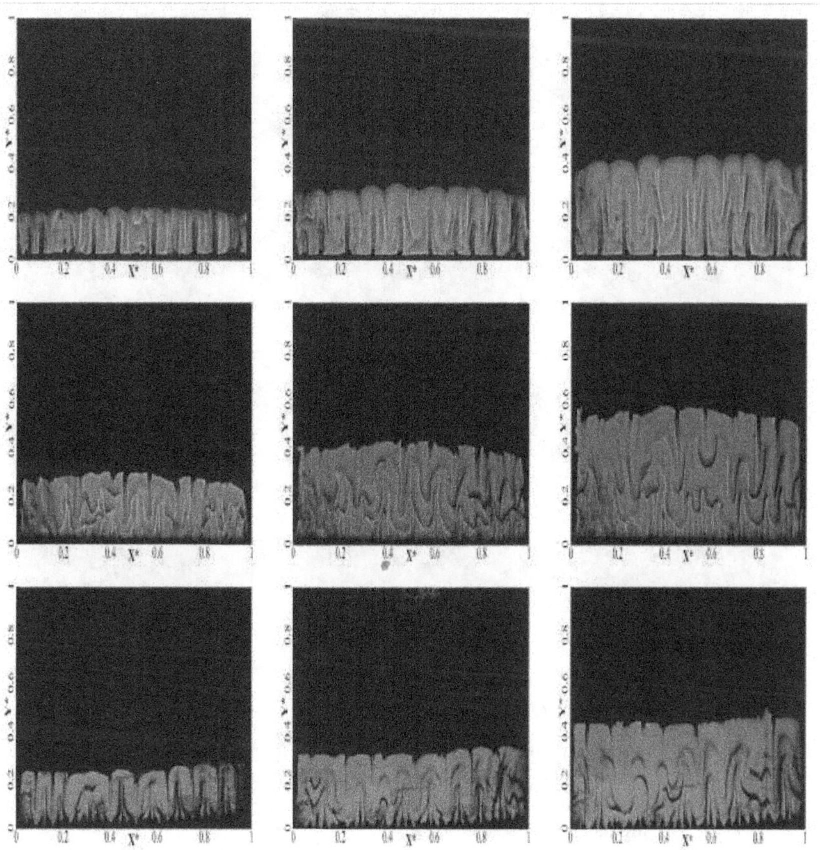

(a) (b) (c)

Figure 6.2: Experimental visualization comparing the evolution of plume structures for three heat-transfer at different time intervals: (a) Fo = 0.0058, (b) Fo = 0.0116, and (c) Fo = 0.0174

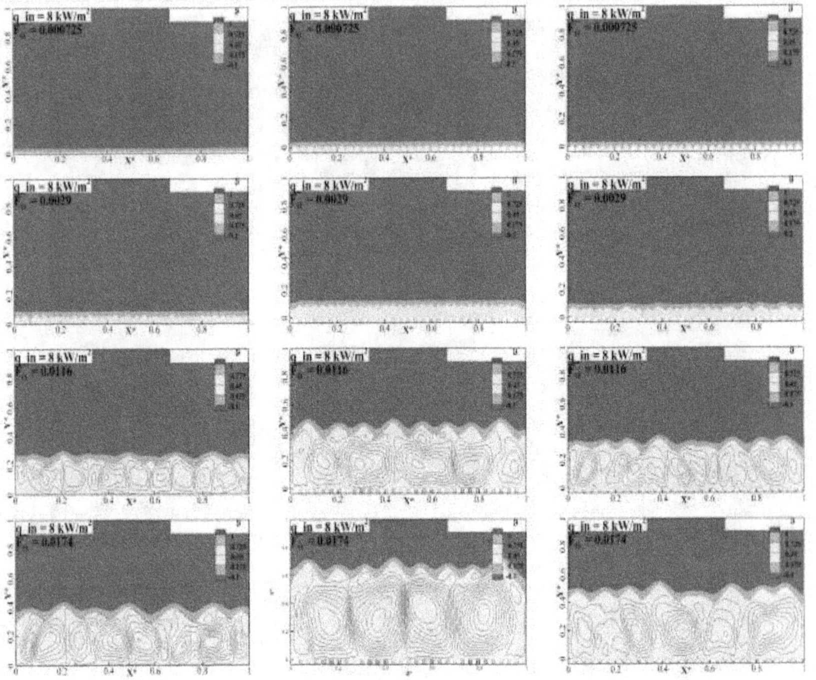

(a) (b) (c)

Figure 6.3: Temporal evolution of coupled thermal and flow fields, shown as temperature contours overlaid with stream function, within a rectangular enclosure at 0° inclination for (a) flat surface, (b) surface with square patterned elements, and (c) surface with triangular patterned elements

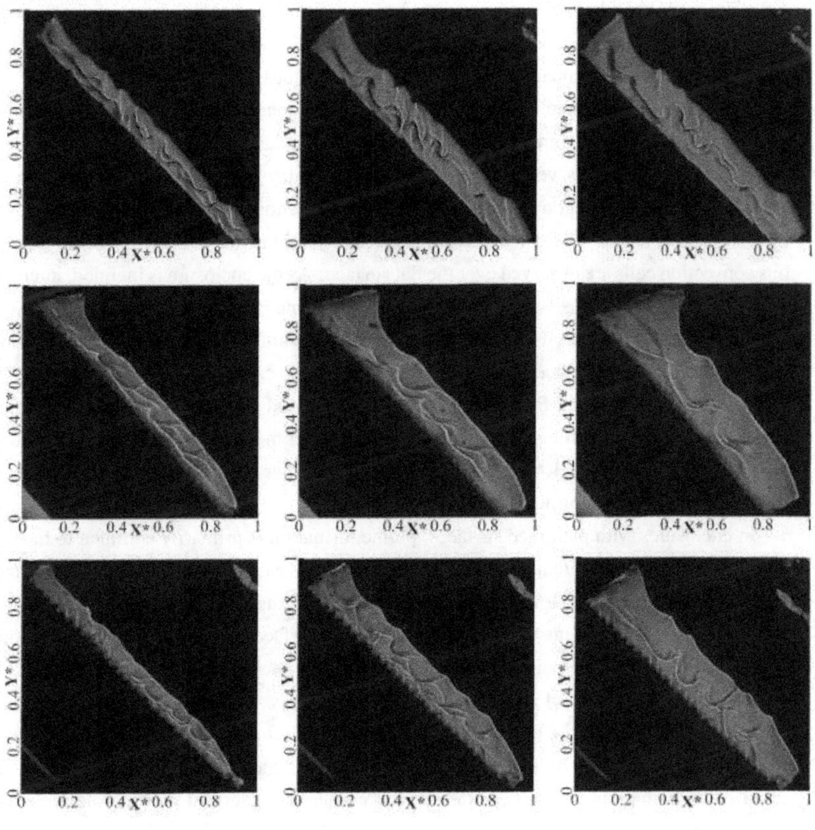

(a) (b) (c)

Figure 6.4: Experimental visualization comparing the evolution of plume structures for three heat-transfer conditions at different time intervals: (a) Fo = 0.0058, (b) Fo = 0.0116, and (c) Fo = 0.0174.

Hot plumes interacting with this dynamic interface introduce discontinuities that enhance local melting. Notably, enhanced melting is observed near the top-left corner of the inclined surface, where the bulk motion of hot melt along the heater surface redirects and intensifies. Similar trends are observed in the corresponding numerical simulations. Figure 6.5 presents the superimposed streamlines and temperature contours for inclined enclosures with flat, triangular, and square patterned surfaces. Tilted latent heat storage systems exhibit markedly different convection dynamics compared to horizontal configurations.

Figure 6.6 presents contours of the y-velocity component obtained from simulations at an identical nondimensional time ($F_O = 0.0058$) under a boundary heat flux of 7 kW/m^2 for various inclination angles. The y-velocity contours highlight buoyancy-induced upward flow and the associated convective heat transfer, with clockwise and anticlockwise circulations represented by positive and negative velocity regions, respectively. In the horizontal configuration, numerous convection cells are observed over the flat surface. As the enclosure is inclined, interactions between these cells intensify, leading to oscillatory plume behavior. At higher inclination angles, distinct convection cells near the heated boundary diminish, as the hot molten PCM tends to flow parallel to the surface. This flow turns near the top corner of the inclined enclosure and impinges on the unmelted PCM, transferring a substantial portion of its thermal energy and thereby accelerating melting in this region. The cooled molten PCM subsequently descends along the phase boundary toward the bottom corner, further contributing to solid PCM erosion through recirculatory motion.

In enclosures with patterned surfaces, plume formation is primarily confined to the roughness elements. The perturbation of convective plumes is reduced due to stronger plume anchoring, which stabilizes the flow even at larger inclination angles. Organized plumes generated over the roughness elements disrupt the oscillatory interface between the hot and cold molten streams. Tilting modifies the convection-cell structure over roughened surfaces; for triangular patterned, the extended surface aligns more favorably to promote organized convection, whereas square patterned elements induce localized recirculation within the cavities. Overall, the combined upward and downward flow layers thicken along the inclined surface, resulting in an enlarged molten region near the top-left corner characterized by multiple recirculation zones.

When the enclosure is inclined, the buoyancy force acting on the molten PCM is no longer aligned vertically with respect to the enclosure coordinates. The heated PCM near the bottom surface becomes less dense and rises along the direction of gravity. As a result, the generated thermal plumes detach from the heated surface and migrate toward the upper region of the inclined enclosure. The inclination introduces a component of the gravitational force parallel to the enclosure surface, which influences the plume trajectory. The hot liquid generated near the heated regions experiences buoyancy-driven acceleration along the inclined direction, causing the plumes to move away from the heating surface and accumulate near the upper portion of the enclosure. This leads to an uneven melting distribution, with greater liquid fraction accumulation in the upper region than in the lower region.

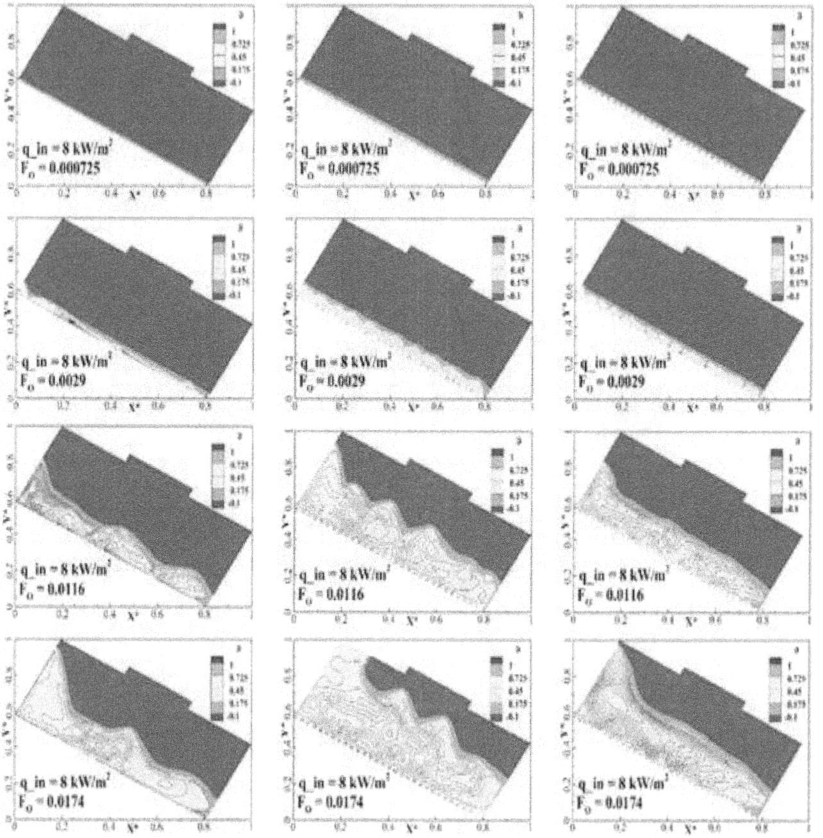

(a) (b) (c)

Figure 6.5: Temporal evolution of the coupled thermal and flow fields, illustrated by temperature contours overlaid with streamlines, within a rectangular enclosure inclined at 30° for (a) a flat surface, (b) a surface with square patterned elements, and (c) a surface with triangular patterned elements.

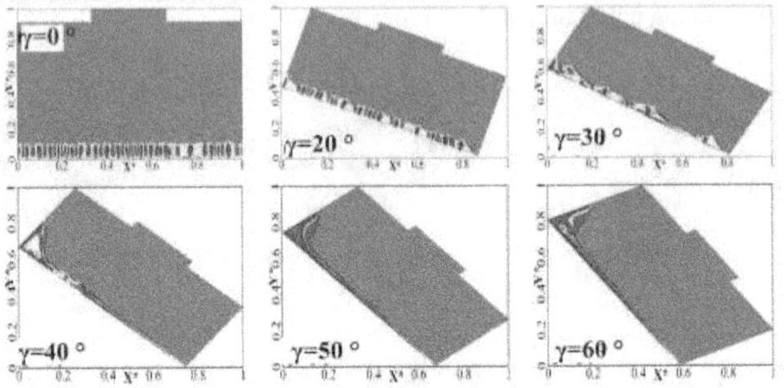

(a) Flat surface

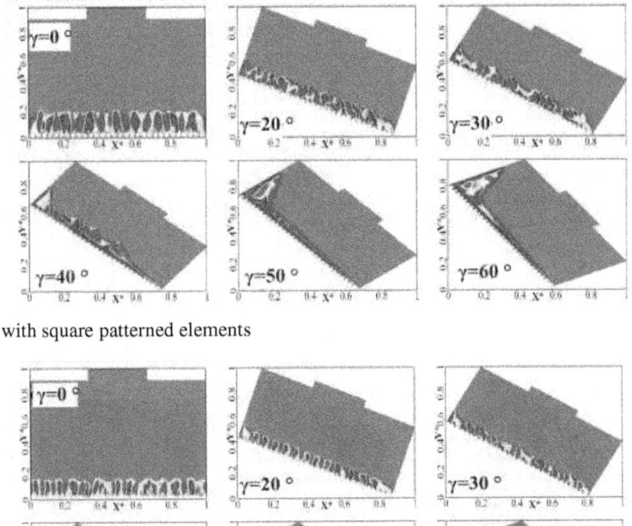

(b) Surface with square patterned elements

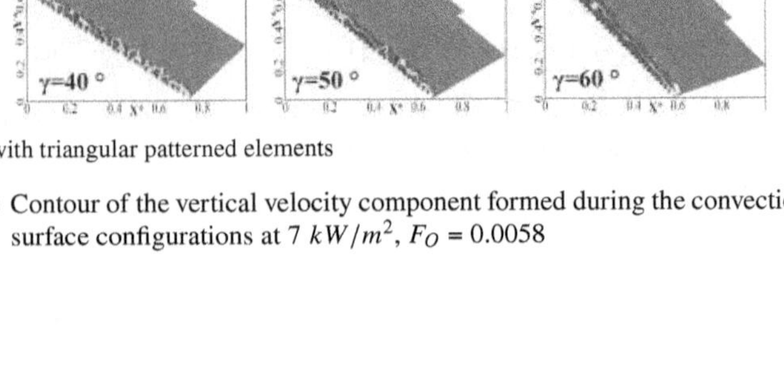

(c) Surface with triangular patterned elements

Figure 6.6: Contour of the vertical velocity component formed during the convection for three surface configurations at 7 kW/m^2, F_O = 0.0058

6.3 Heat transfer augmentation due to patterned surfaces for inclined latent heat storages

Previous studies have established the advantages of extended surfaces in enhancing melting when the enclosure is maintained in a horizontal orientation, primarily due to the increased effective heat-transfer area provided by surface roughness. However, changing the orientation of the enclosure introduces additional geometric and physical effects that can significantly alter melting behavior. Investigating these effects constitutes a primary objective of the present study. Accordingly, a comprehensive analysis is conducted to evaluate melt fraction evolution in PCM enclosures heated by a flat surface, a surface with square patterned elements, and a surface with triangular patterned elements over a range of inclinations ($0° \leq \sigma \leq 60°$) and applied heat fluxes (7-10 kW/m^2). Initially, the melt fraction progression for the three surface configurations at different inclinations is compared under a fixed moderate heat flux of 8 kW/m^2. As shown in Fig.6.7a, the melt fraction consistently decreases with increasing inclination for all surface types. Among the configurations, the surface with square patterned elements exhibits the highest melt fraction across most orientations, reflecting the enhanced heat-transfer capability of extended surfaces. However, at later stages of melting and higher inclination angles, this advantage diminishes due to the increased flow resistance imposed by square patterned elements, particularly when the phase interface remains close to the heating surface.

The surface with triangular patterned elements displays a more linear and consistent melt fraction progression across all orientations, as its geometry offers less obstruction to melt flow. Moreover, the inclined face of each triangular element promotes the generation of upward convection currents as it aligns more favorably with the direction of gravity. At higher heat flux levels, surfaces with square patterned elements again demonstrate superior melting performance, as the melt front moves away from the heating surface, providing sufficient space for the molten PCM to rise along the extended surfaces. A comprehensive comparison of melt fraction results for all surface configurations, inclination angles, and heat flux levels investigated in this study is presented in Fig. 6.8 .

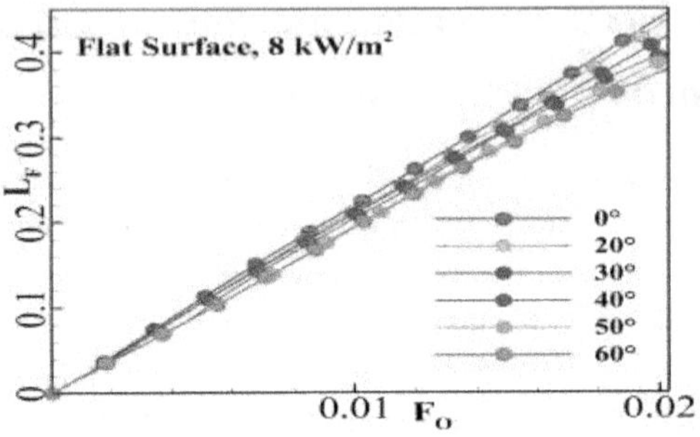

(a) Flat surface

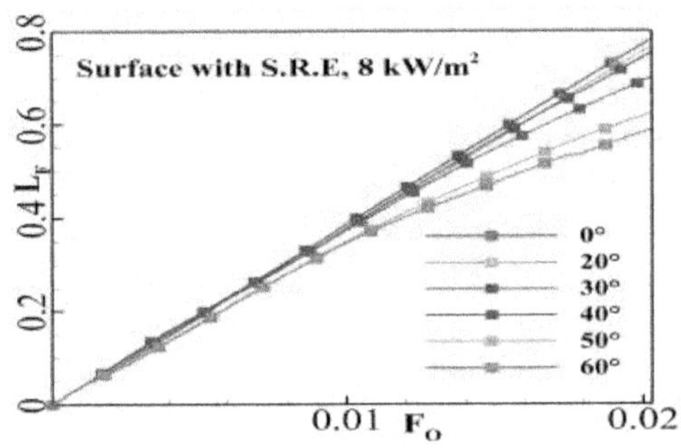

(b) Surface with square patterned elements

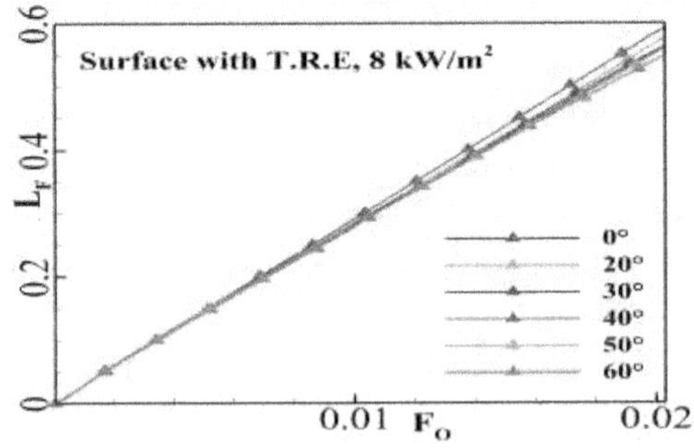

(c) Surface with triangular patterned elements

Figure 6.7: Liquid fraction comparison for (a) flat surface, (b) surface with square patterned elements, and (c) surface with triangular patterned elements at various orientations ($0° \leq \sigma \leq 60°$) under a fixed heat flux of 8 kW/m^2.

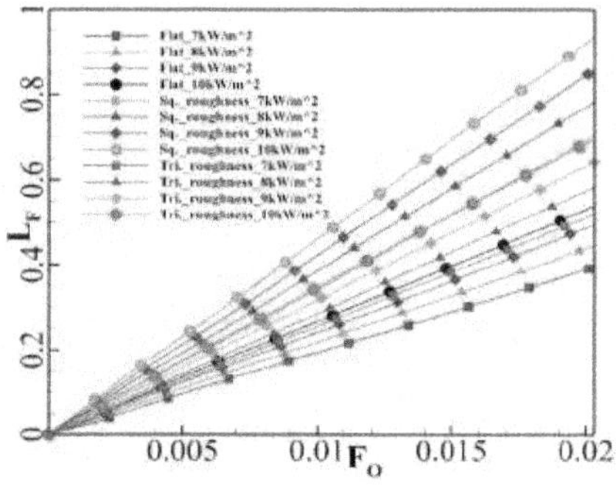
0.2
0.4
0.6
0.8
0.005
0.01
0.015
0.02

(a) 0°

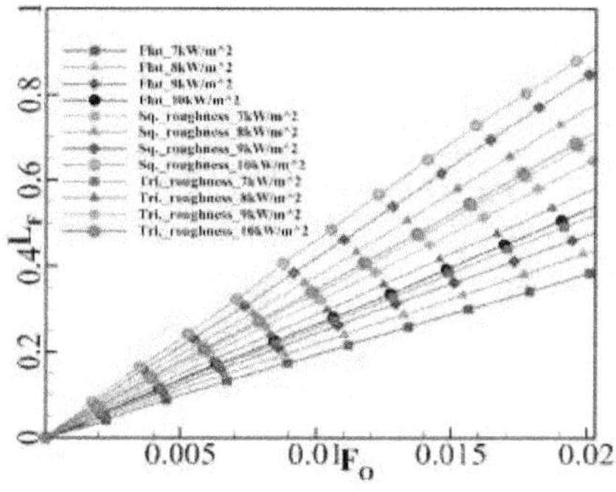
0.2
0.4
0.6
0.8
0.005
0.01
0.015
0.02

(b) 20°

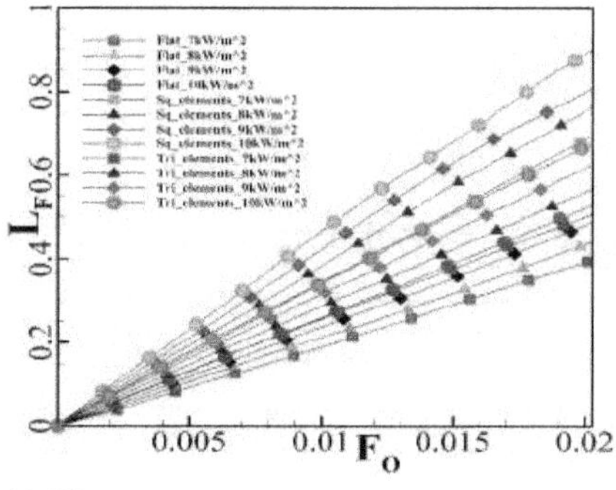
0.2
0.4
0.6
0.8
0.005
0.01
0.015
0.02

(c) 30°

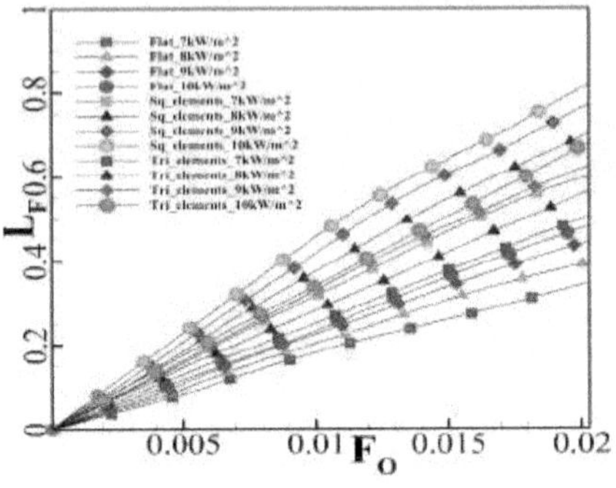

(d) 40°

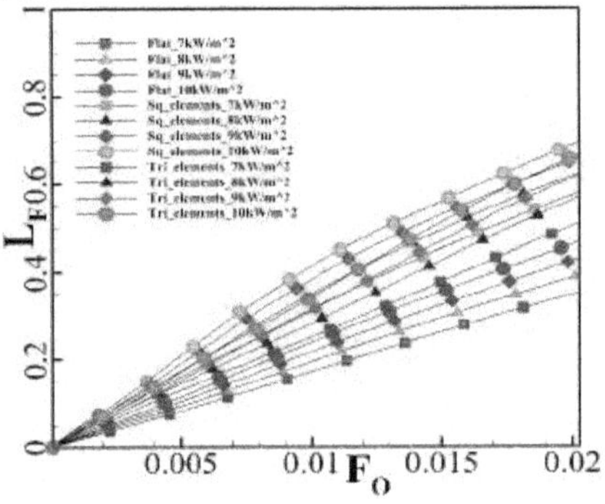

(e) 50°

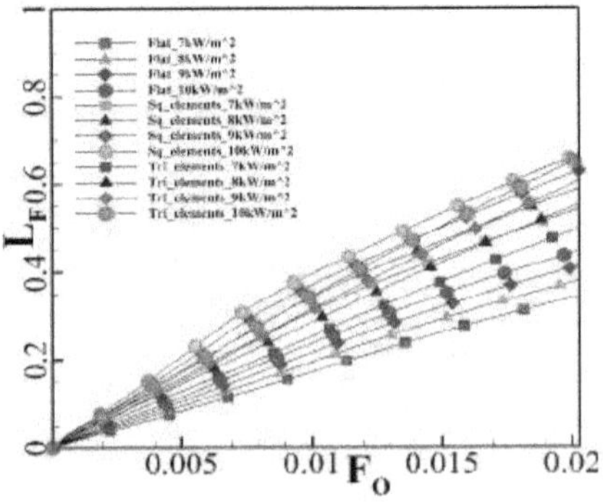

(f) 60°

Figure 6.8: Liquid fraction comparison for flat, square-patterned, and triangular-patterned surfaces across different orientations ($0° \leq \gamma \leq 60°$) and surface heat fluxes (7–10 kW/m^2).

6.4 Heat transfer correlations for inclined latent heat storages with patterned heating surfaces

The preceding discussion clearly demonstrates that inclined PCM enclosures with patterned heating surfaces exhibit distinct melting and heat-transfer characteristics. The combined effects of enclosure inclination and surface roughness lead to heat-transfer behavior that differs significantly from that of conventional horizontal configurations. To quantify these effects, a detailed analysis of instantaneous and time-averaged Nusselt numbers is conducted for PCM enclosures heated using a flat surface, a surface with square patterned elements, and a surface with triangular patterned elements over a range of orientations ($0° \leq \gamma \leq 60°$) and applied heat fluxes (7-10 kW/m^2).

Figure 6.9 presents the temporal evolution of the instantaneous Nusselt number for the three surface configurations at four inclination angles under a surface heat flux of 7 kW/m^2. During the initial melting stage, heat transfer is primarily governed by diffusion, resulting in relatively low Nusselt numbers. In this regime, patterned surfaces exhibit slightly higher heat transfer owing to their increased effective surface area. Following the onset of thermal instability, the heat-transfer rate increases steadily with the strengthening of buoyancy-driven convection. At zero and small inclination angles, all surface configurations display comparable heat-transfer rates during the convection-cell coarsening regime. As the inclination increases to moderate levels, the buoyancy component parallel to the heating surface becomes more pronounced, disrupting the regular convection-cell structure. The plume oscillations observed at 30° inclination (Fig. 6.5) are similar across all surface types, leading to nearly identical heat-transfer rates at larger melt heights.

As melting progresses and convection cells grow, patterned surfaces provide enhanced heat transfer at horizontal and low-inclination orientations. However, with increasing inclination, the lower portion of the inclined surface becomes less effective due to significant non-uniformity in melt height—being greatest near the top-left corner and minimal near the bottom-right corner of the enclosure. Consequently, a larger pool of molten PCM accumulates near the upper region, and all surface configurations tend to exhibit similar instantaneous Nusselt numbers toward the later stages of melting at higher inclination angles. Additionally, the present study reveals that the heat-transfer performance of square patterned elements deteriorates as melting advances, owing to molten PCM entrapment near the surface, which increases flow resistance. The time-averaged Nusselt numbers at various inclination angles, shown in Fig. 6.10, indicate that surfaces with triangular patterned elements consistently deliver superior heat-transfer performance across all tilted configurations and applied heat fluxes. While square patterned elements are more effective in the horizontal orientation due to their larger surface area, the cavities between these elements become less efficient in supporting convective heat transfer under inclined conditions. In contrast, triangular patterned elements offer significantly enhanced heat transfer at higher inclinations, as their inclined lateral surfaces direct heat flux more favorably along the direction of gravity, thereby promoting vertical convection.

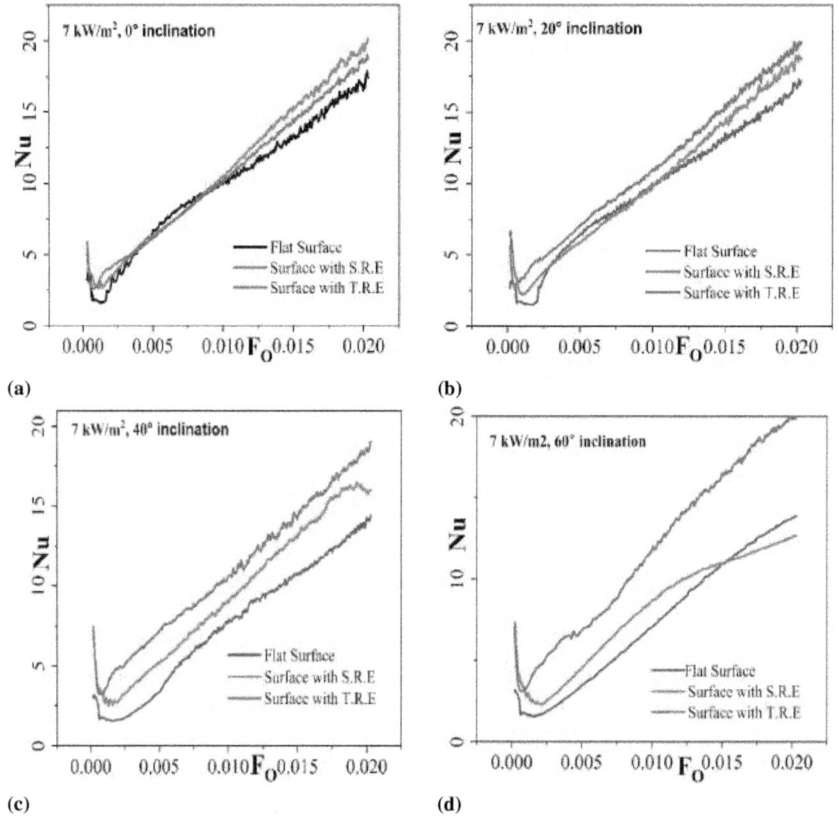

Figure 6.9: Instantaneous Nusselt number evolution with time for the three heat-transfer surface configurations at different inclination angles.

Notably, at an inclination of 30°, the orientation of the triangular patterned elements becomes particularly advantageous, resulting in maximum heat-transfer enhancement. Furthermore, these surfaces impose less resistance to the upward movement of bulk molten PCM near the heating surface, especially at larger inclination angles.

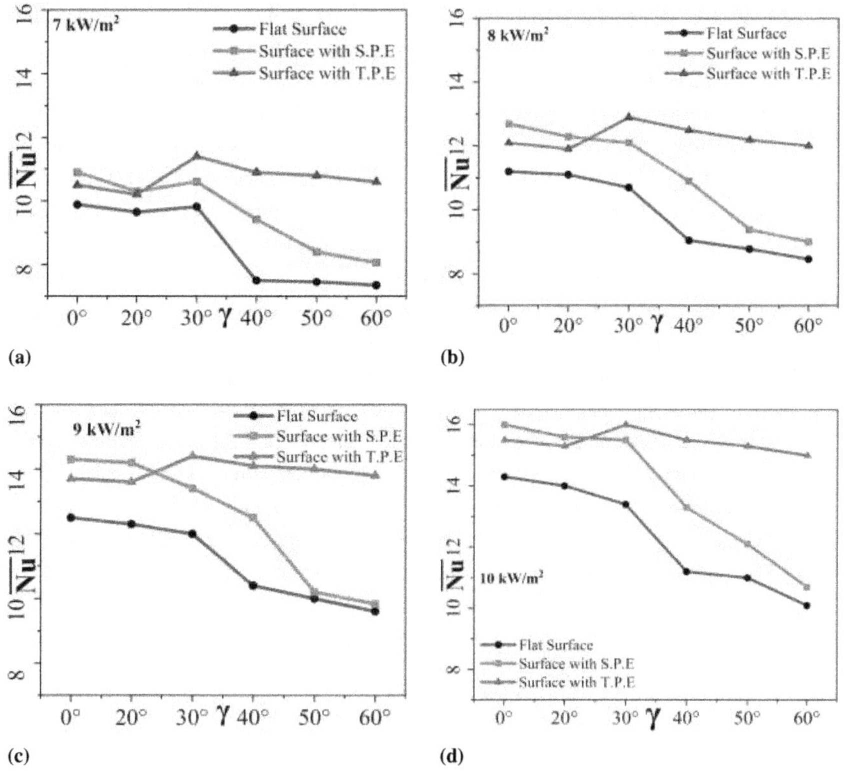

Figure 6.10: Time-averaged Nusselt numbers at different inclination angles for the three heat-transfer surface configurations.

6.5 Scaling laws and heat transfer correlations for inclined PCM confinements with patterned heating surfaces

Most of the correlations available in the literature have been developed for melting over flat heating surfaces oriented horizontally. In the present study, scaling laws and heat-transfer correlations are formulated for inclined PCM enclosures with patterned heating surfaces based on an extensive set of heat-transfer data. Classical correlations, such as the Prandtl–Blasius theory, report a Rayleigh number exponent of 0.25, while Madruga and Curbelo [[93]] obtained a higher value of 0.29. Similar investigations have also been reported for semicircular cavities with straight and wavy heating surfaces under constant heat-input conditions.

In the current work, PCM enclosures with patterned heating surfaces subjected to constant heat fluxes ranging from 7 to 10 kW/m^2 and oriented at various inclination angles are considered to develop generalized correlations. Following standard practice, the Rayleigh number formulation incorporates the $g \cos \gamma$ term to account for the component of gravitational acceleration normal to the inclined surface, which governs buoyancy-driven flow. Figure 6.11 illustrates the evolution of the Nusselt number as a function of the Rayleigh number for inclined PCM enclosures with patterned heating surfaces during the melting process. The corresponding coefficients of determination (R^2) are summarized in Table 1. The Rayleigh number exponent in the proposed correlations varies between 0.19 and 0.22 for configurations with smaller inclination angles, which is close to conventional values reported in the literature. These correlations are expected to be useful for the design and development of latent heat energy storage systems and PCM-based cooling applications that must operate under inclined mounting conditions due to practical system constraints.

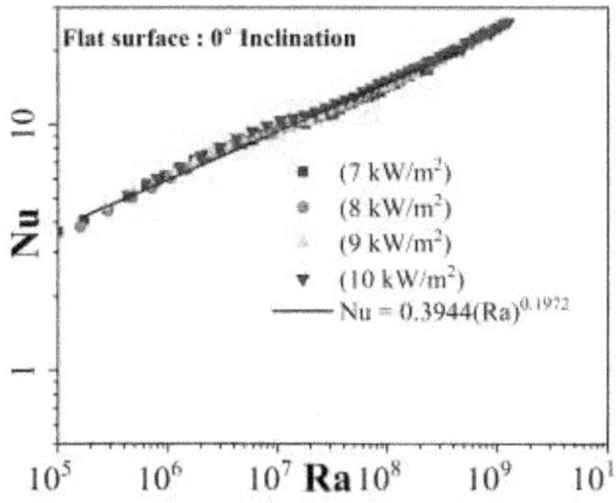

(**a**) Flat surface (0°)

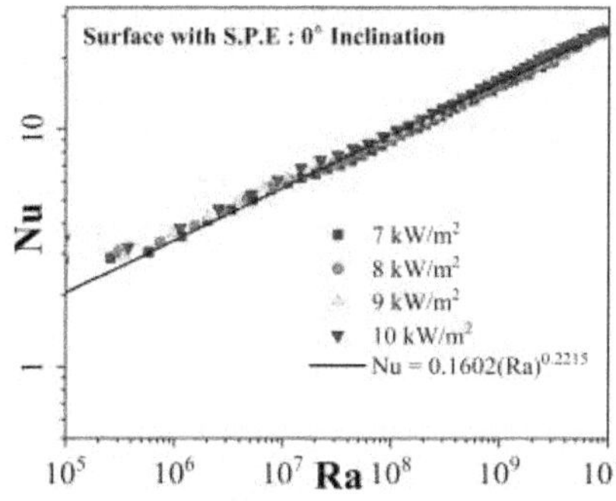

(**b**) Square-patterned surface (0°)

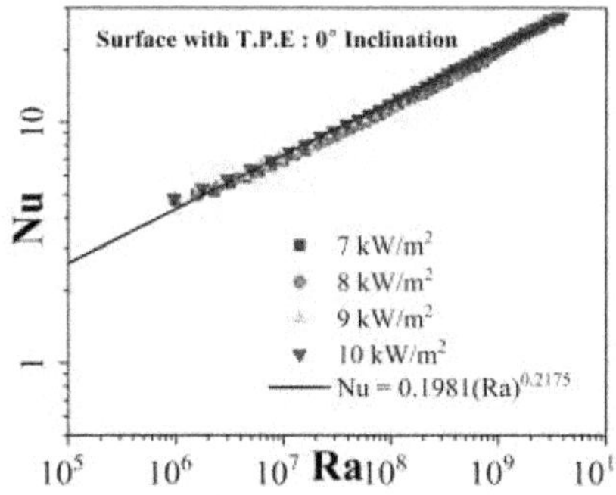

(**c**) Triangular-patterned surface (0°)

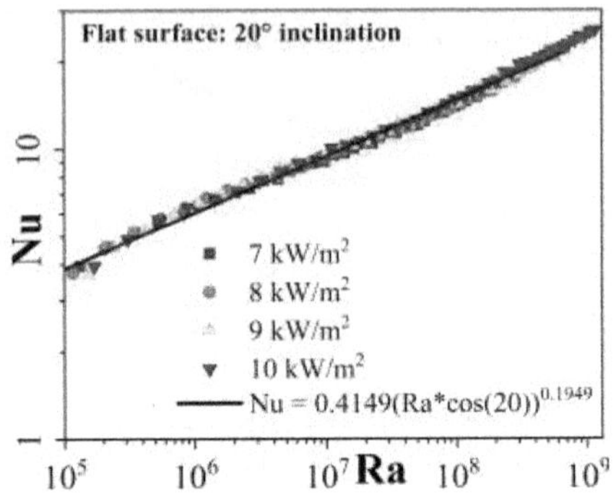

(d) Flat surface (20°)

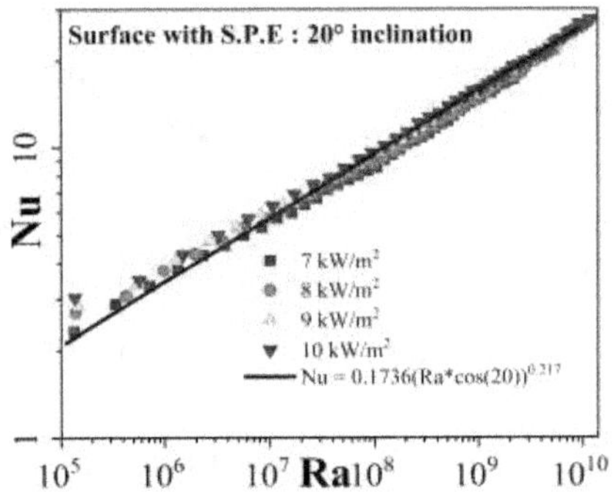

(e) Square-patterned surface (20°)

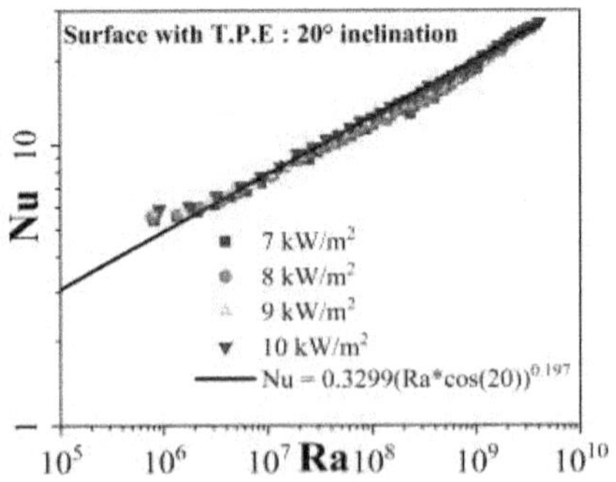

(f) Triangular-patterned surface (20°)

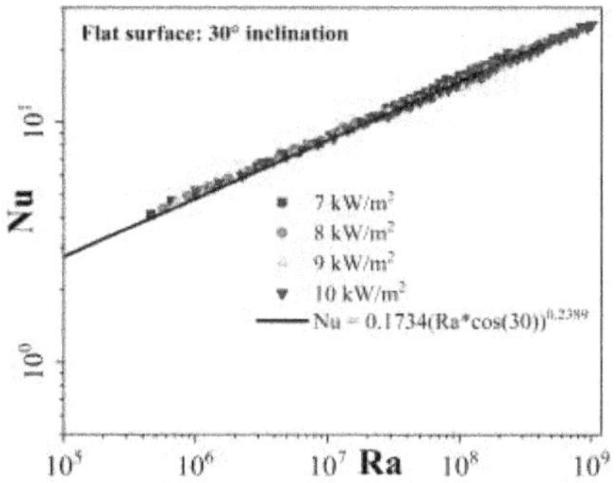

(g) Flat surface (30°)

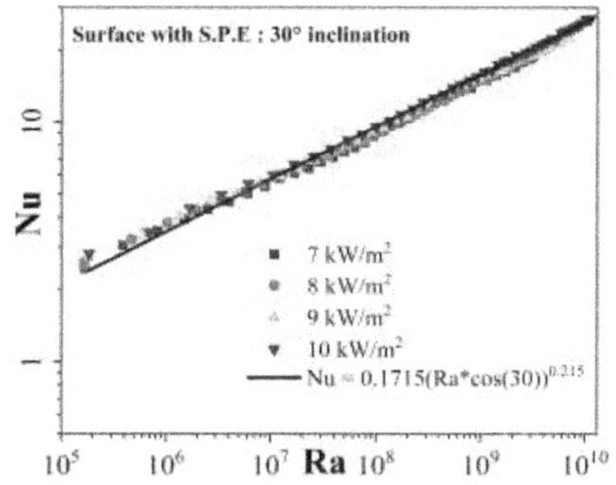

(h) Square-patterned surface (30°)

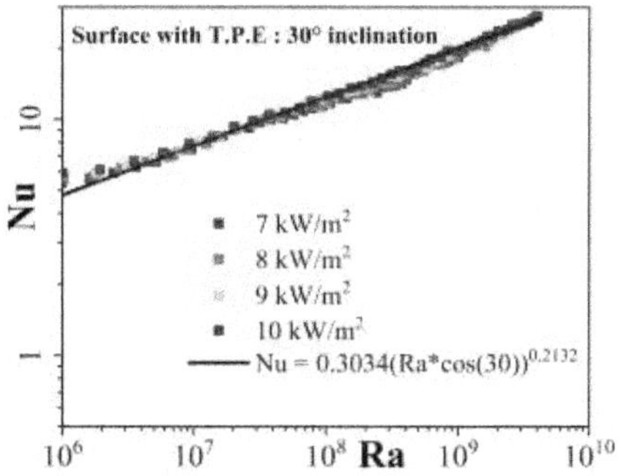

(i) Triangular-patterned surface (30°)

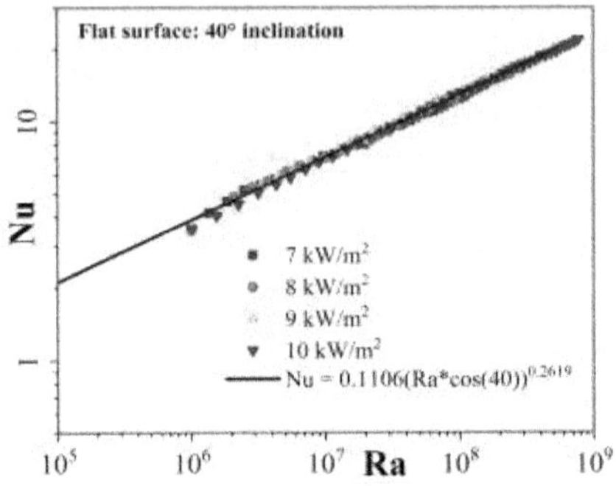

(j) Flat surface (40°)

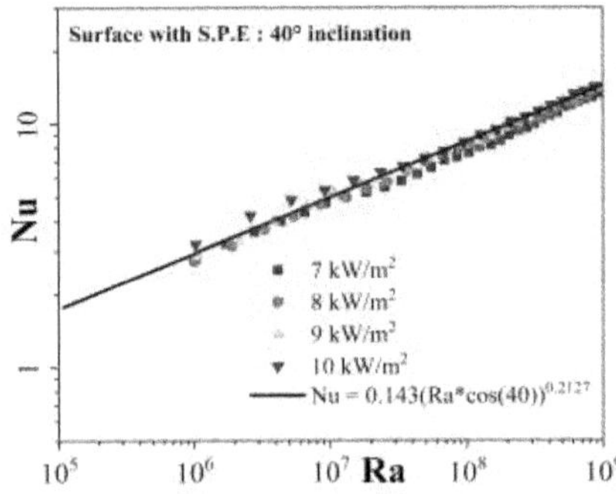

(k) Square-patterned surface (40°)

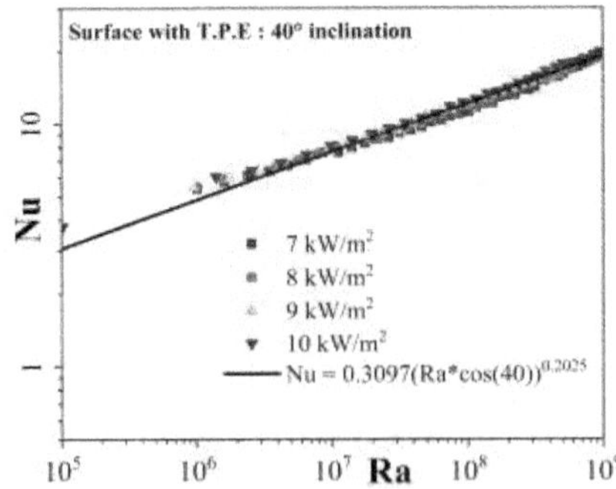

(l) Triangular-patterned surface (40°)

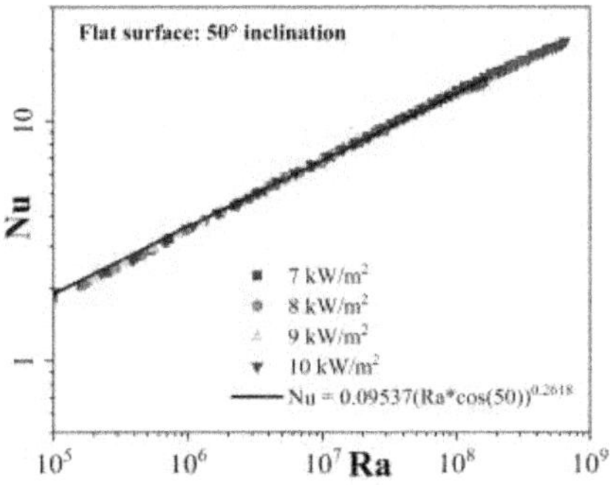

(**m**) Flat surface (50°)

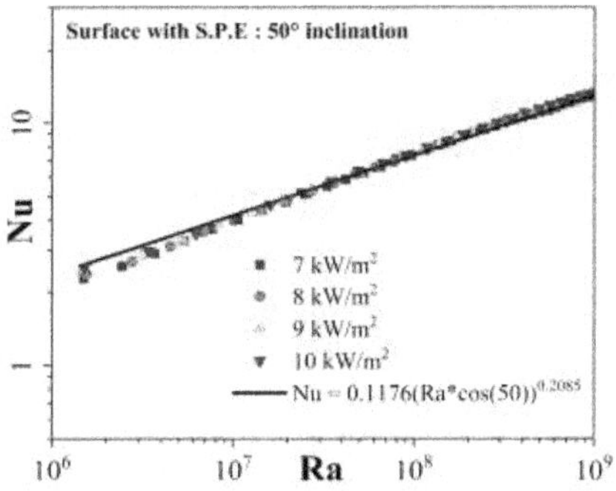

(**n**) Square-patterned surface (50°)

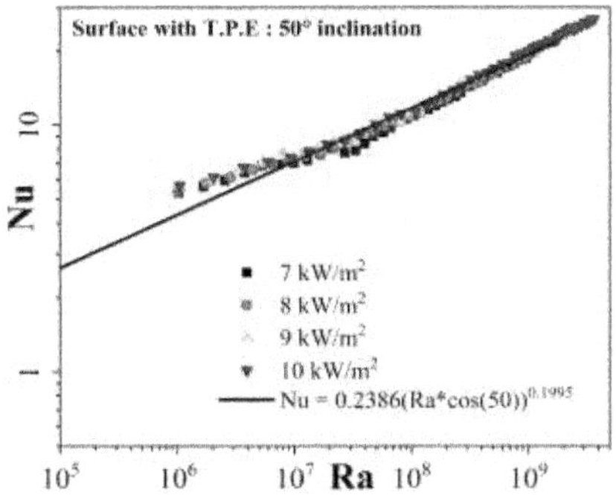

(**o**) Triangular-patterned surface (50°)

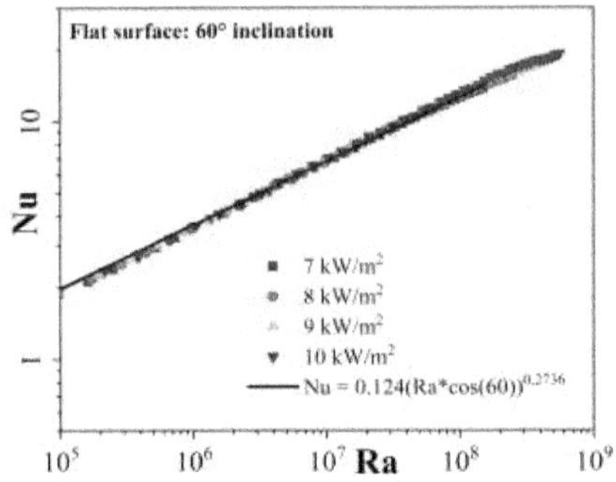

(p) Flat surface (60°)

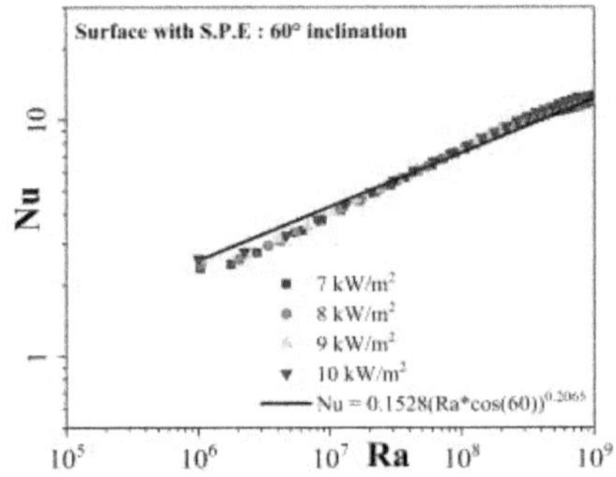

(q) Square-patterned surface (60°)

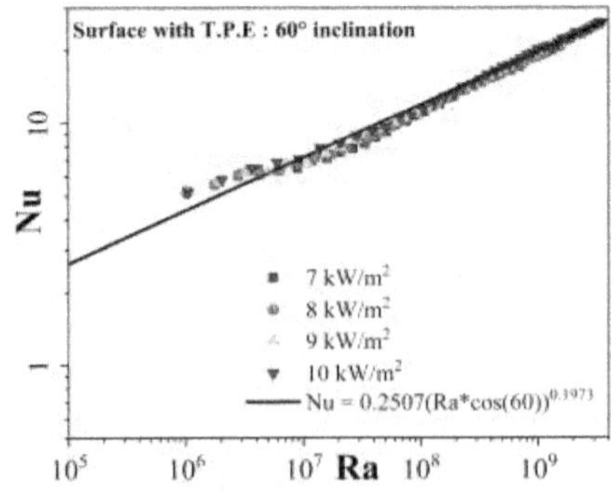

(r) Triangular-patterned surface (60°)

Figure 6.11: Correlations of Nusselt number (Nu) with Rayleigh number (Ra) for inclined PCM confinements with flat, Square-patterned, and Triangular-patterned heating surfaces at inclination angles ranging from 0° to 60°.

6.6 Summary

This work presents combined experimental and numerical investigations of PCM melting in an inclined rectangular enclosure with flat, square and triangular-patterned bottom surfaces under constant heat flux. Heat transfer characteristics are analysed through Nusselt-Rayleigh scaling for various inclinations (0°-60°) and heat fluxes. At 0° inclination, the square-patterned surface yields the highest Nu-Ra exponent, whereas beyond 30° the flat surface shows superior scaling. Overall, square roughness provides the highest melting rate for all inclinations, while triangular patterned surface becomes most effective at 30°, producing the maximum time-averaged Nusselt number.

7 Conclusion

Performance enhancement of latent heat storage is an active research theme over past few decades owing to the increasing global energy demand and limited supply highlight the urgent need for the development of highly efficient energy storage technologies. Design improvements of latent energy storage systems are essential for optimizing faster phase transitions in applications requiring rapid energy delivery. Hence the primary objective of the present work is to investigate the effects of different heating surface configurations, orientations, and boundary conditions on the melting behavior of a phase change material (PCM) for latent heat energy storage applications. In addition to the experimental investigations, numerical simulations are performed using the conventional Navier-Stokes equations along with the enthalpy-porosity method. The insights obtained from this combined approach enable a deeper understanding of several important melting and heat transfer mechanisms that have not been extensively explored in earlier studies. This chapter summarizes the key findings of the present research and also provides recommendations for future work to further enhance and extend the understanding of heat transfer and melting dynamics in latent heat energy storage systems.

7.1 Major contributions based on the present work

- An experimental facility for the visualization melt front development and thermal convection in PCM containers of various types of heating surfaces and orientations is realized.
- High fidelity temperature responses at various locations inside the PCM confinement are grabbed for numerical simulation procedure validation.
- An enthalpy-porosity approach based finite volume computational procedure is developed for the numerical study.
- The developed computational procedure is well validated with in-house experimental measurements.
- Study elucidates the melting evolution, convective cell evolution at various levels of boundary flux for the three variations of bottom surface geometry and is extended to quantify the heat transfer coefficient for various levels of boundary flux provided through different types of heating element surfaces.

- Heat transfer correlations are developed for melting over various kinds of heater surfaces, heat flux conditions, and orientations of the confinement.

7.2 Important conclusions from the current study

Experimental and numerical investigations on melting dynamics and heat transfer have been carried out for PCM confinements with various types of heater surfaces, heating patterns and confinement orientations. Following are the specific conclusions from the study.

- Studies for the semi-circular PCM confinement indicates that the average Nusselt number is higher in case of heating with flat surface compared with that of wavy and corrugated surface, the least occur in case of corrugated surface case. Correlations of the form Nu=a$\cdot Ra^n$ for the Nusselt number evolution with Rayleigh number is developed for all the test cases experimented and estimated correlation coefficients. Evolution of the non-dimensional boundary layer thickness with respect to Rayleigh number is estimated and correlated in the form $\delta_T \approx Ra^{-0.2}$ which is unique the physics of transport process of melting in semi-circular enclosure.

- Studies on melting phase in a PCM heated with wavy heating boundary having boundary heat fluctuations of time periods comparable with that of the time scale of Rayleigh–Bénard convection cells has been carried out. This evidences that the convection cells during melt advancement over wavy heating surfaces with low AWR create strong convection currents during the early stages of melting. However, as time elapse, this effect decreasing as the height of the melt increases. This study establishes a strong dependence of amplitude of surface temperature fluctuations in time-periods of convection current. Moreover, wavy surfaces with high AWR produce intense convection plumes that attach to the crest of the waves as melting progresses, leading to the growth of convection cells. This study reveals a non-linear melt fraction evolution with respect to time for wavy boundaries subjected to fluctuating heat input in contrary to the linear variation melt fraction development with respect to time for a steady temperature boundary condition. A significant enhancement in melt fraction is noted as the amplitude of boundary temperature variation increases, regardless of the degree of surface waviness.

- Typical power law relations such as power-law relationships-such as $L_F \sim F_O^m$, h*$\sim F_O^p$ and Nu~Ra applicable for melt development over wavy heating surface with time-varying heat inputs are established for the first time in the current study which agrees well with established relations for base limiting cases for flat surfaces.

- Studies on the effect of inclination of PCM confinement exhibits a unique convection plume pattern forming a wavy interface between hot and cold region of the melt pool. Melting is found to enhance at the top portion of the inclined confinement due to the interaction of the molten PCM owing to the buoyancy. Square surface pattern realizes more melting when the PCM confinement is inclined. A correlation of the form Nu~ $(Ra\cos\theta)^m$ has been developed for the estimation of heat transfer from patterned heating surfaces of inclined PCM encapsulations. Exponents close to the well-established correlations (0.2) applicable for horizontal encapsulations with flat heating surfaces have also been observed in the current study. The Nu Vs Ra exponent ranges from 0.20 to 0.22 for the surface with square roughness elements and from 0.19 to 0.21 for the surface with triangular roughness elements.

Observations on melting dynamics and estimated correlations heat transfer from wavy as well as patterned heating surfaces and inclined PCM encapsulations are highly useful in the design of latent heat storages and PCM based cooling systems.

7.3 Future Scope

Although extensive research has been carried out on melting dynamics using roughened heating surfaces, the present experimental flow visualization is performed using a shadowgraph setup. To obtain detailed and quantitative information on the flow field, particularly in the vicinity of roughened or wavy walls, future investigations shall employ Particle Image Velocimetry (PIV) techniques for more accurate flow visualization and velocity measurement. Future work may involve conducting both experimental and numerical investigations incorporating microsized particles to examine their influence on melting dynamics and heat transfer characteristics. Investigations may be performed for finding the effect of surface roughness on the melting of a phase-changing material under thermocapillary effects. Furthermore, future investigations may also focus on enhancing the thermal performance of phase changing materials through the application of magnetohydrodynamic (MHD) effects.